CAPACITOR BASICS

FIRST EDITION

BY PRASUN BARUA

ABOUT

Welcome to Capacitor Basics! This is a nonfiction science book which contains various topics on basics of Capacitor. A capacitor is a two-terminal electrical device which uses an electric charge to store energy. It is made up of two electrical conductors separated by a distance. The space between the conductors can be filled with a vacuum or a dielectric, which is an insulating material. Capacitance refers to the capacitor's ability to store charges. Capacitors store energy by keeping opposing charges apart. A parallel plate capacitor is the most basic type of capacitor, consisting of two metal plates with a gap between them. Capacitors come in a wide variety of shapes, sizes, lengths, girths, and materials. The ratio of a system's change in electric charge to the corresponding change in its electric potential is known as capacitance. The capacitance of any capacitor can be either fixed or variable, depending on its usage. Apart from charge and voltage, capacitance also depends on the shape and size of the capacitor and also on the insulator used between the conducting plates. Capacitors are used for storing energy, power conditioning and signal processing. This is the first edition of the book. Thanks for reading the book.

CONTENTS

CHAPTER-1

INTRODUCTION TO CAPACITOR

When connected to a voltage source, a capacitor is a simple passive device capable of storing an electrical charge on its plates. The capacitor is a component that, like a small rechargeable battery, has the ability or "capacity" to store energy in the form of an electrical charge that produces a potential difference (Static Voltage) across its plates. Capacitors come in a variety of sizes and shapes, ranging from very small capacitor beads used in resonance circuits to large power factor correction capacitors, but they all do the same thing: they store charge.

A capacitor, in its most basic form, is consisting of two or more parallel conductive (metal) plates that are not connected or touching each other, but are electrically separated by air or some form of good insulating material. This insulating material could be waxed paper, mica, ceramic, plastic, or a liquid gel similar to that found in electrolytic capacitors. It is worth noting that the insulating layer between a capacitor's plates is commonly referred to as the Dielectric.

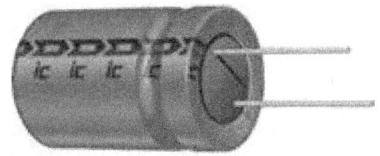

A Typical Capacitor

This insulating layer prevents DC current from flowing through the capacitor, instead allowing a voltage to be present across the plates in the form of an electrical charge.

A capacitor's conductive metal plates can be square, circular, or rectangular, or cylindrical or spherical, with the general shape, size, and construction of a parallel plate capacitor depending on its application and voltage rating.

When used in a direct current (DC) circuit, a capacitor charges up to its supply voltage but blocks current flow because the dielectric of a capacitor is non-conductive and thus acts as an insulator. When a capacitor is connected to an alternating current (AC) circuit, the current appears to flow through the capacitor with little or no resistance.

A positive charge in the form of Protons and a negative charge in the form of Electrons are the two types of electrical charge.

When a direct current (DC) voltage is applied to a capacitor, the positive (+ve) charge quickly accumulates on one plate while the opposite negative (-ve) charge quickly accumulates on the other.

A charge of the same sign will depart from the -ve plate for every particle of +ve charge that arrives at one plate. The plates then remain charge neutral, and a potential difference between the two plates is established as a result of this charge.

Due to the insulating properties of the dielectric used to separate the plates, electrical current cannot flow through the capacitor and around the circuit once it reaches its steady state condition.

The flow of electrons onto the plates is referred to as the capacitor's Charging Current, and it continues until the voltage across both plates (and thus the capacitor) equals the applied voltage Vc.

The capacitor is said to be "fully charged" with electrons at this point. When the plates are fully discharged (initial condition), the strength or rate of this charging current is at its maximum and gradually decreases to zero as the plates charge up to a potential difference across the capacitor's plates equal to the source voltage.

The amount of potential difference present across the capacitor is determined by how much charge was deposited onto the plates by the work done by the source voltage as well as the capacitance of the capacitor, as illustrated below.

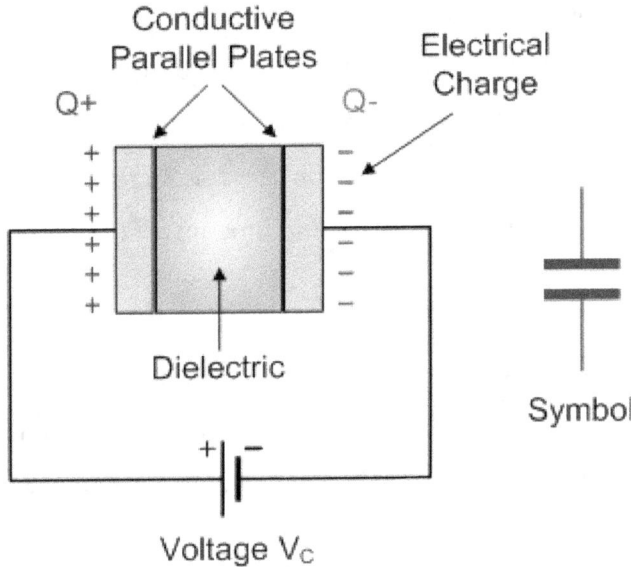

The parallel plate capacitor is the most fundamental type of capacitor. It can be built with two metal or metallized foil plates parallel to each other, with the capacitance value in Farads determined by the surface area of the conductive plates and the distance between them.

Changing any two of these values changes the capacitance value, which is the basis for variable capacitor operation. Furthermore, since capacitors store the energy of electrons in the form of an electrical charge on the plates, the greater the charge that the capacitor holds for any given voltage across its plates, the larger the plates and/or smaller their separation.

In other words, larger plates result in a shorter distance, more capacitance. When a voltage is applied to a capacitor and the charge on the plates is measured, the ratio of the charge Q to the voltage V gives the capacitance value of the capacitor, which is given as C = Q/V. This equation can also be rearranged to yield the well-known formula for the amount of charge on the plates, Q = C x V.

Although the charge is stored on the plates of a capacitor, the energy within the charge is actually stored in an "electrostatic field" between the two plates. When an electric current flows through the capacitor, it charges up, causing the electrostatic field to become much stronger as more energy is stored between the plates.

Similarly, as the current flows out of the capacitor, discharging it, the potential difference between the two plates decreases, as does the electrostatic field, as energy moves out of the plates. The capacitance of a capacitor is its ability to store charge on its plates in the form of an electrostatic field. Not only that, but capacitance is the property of a capacitor that resists voltage changes across it.

The Capacitance of a Capacitor

Capacitance is an electrical property of a capacitor which measures its ability to store an electrical charge on its two plates. The Farad (abbreviated to F) is the unit of capacitance named after the British physicist Michael Faraday. Capacitance is defined as the capacitance of a capacitor when a charge of one coulomb is stored on the plates by a voltage of one volt.

It is important to note that capacitance, C, is always positive in value and has no negative units. However, because the Farad is a very large unit of measurement on its own, sub-multiples of the Farad are commonly used, such as micro-farads, nano-farads, and pico-farads.

Standard Units of Capacitance

- **Microfarad (μF) 1μF = 1/1,000,000 = 0.000001 = 10^{-6} F**
- **Nanofarad (nF) 1nF = 1/1,000,000,000 = 0.000000001 = 10^{-9} F**
- **Picofarad (pF) 1pF = 1/1,000,000,000,000 = 0.000000000001 = 10^{-12} F**

Then using the information above we can construct a simple table to help us convert between pico-Farad (pF), to nano-Farad (nF), to micro-Farad (μF) and to Farads (F) as shown.

Pico-Farad (pF)	Nano-Farad (nF)	Micro-Farad (µF)	Farads (F)
1,000	1.0	0.001	
10,000	10.0	0.01	
1,000,000	1,000	1.0	
	10,000	10.0	
	100,000	100	
	1,000,000	1,000	0.001
		10,000	0.01
		100,000	0.1
		1,000,000	1.0

Capacitance

A parallel plate capacitor's capacitance is proportional to the area, A in meters^2, of the smallest of the two plates and inversely proportional to the distance, d (i.e., the dielectric thickness) given in meters between these two conductive plates.

The generalized equation for a parallel plate capacitor's capacitance is as follows: $C = \varepsilon(A/d)$ where ε represents the absolute permittivity of the dielectric material being used. The dielectric constant, ε_o also known as the "permittivity of free space" has the value of the constant 8.854×10^{-12} Farads per meter.

To make the math a little bit simpler, this dielectric constant of free space, ε_o, which can be written as: $1/(4\pi \times 9 \times 10^9)$, may also have the units of picofarads (pF) per meter as the constant giving: 8.85 for the value of free space. Note though that the resulting capacitance value will be in picofarads and not in farads.

Instead of a perfect vacuum, the conductive plates of a capacitor typically have some sort of insulating material or gel between them. We can use the permittivity of air, and particularly of dry air, as being very close to the same value as a vacuum when calculating the capacitance of a capacitor.

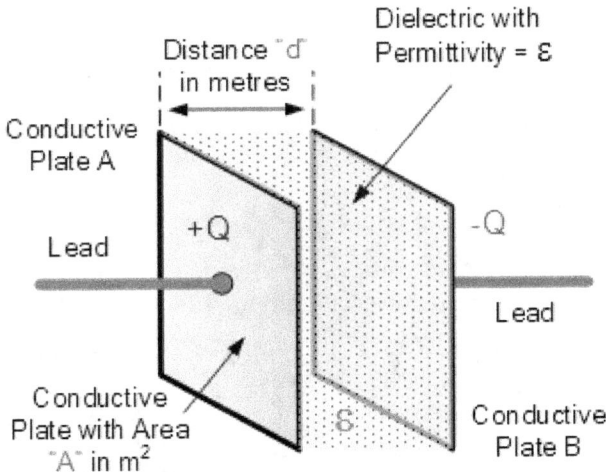

Example No1

A capacitor is constructed from two conductive metal plates 30cm x 50cm which are spaced 6mm apart from each other, and uses dry air as its only dielectric material. Calculate the capacitance of the capacitor.

Using: $C = \varepsilon_o \dfrac{A}{d}$

where: $\varepsilon_o = 8.854 \times 10^{-12}$

$A = 0.3 \times 0.5\ m^2$ and $d = 6 \times 10^{-3}\ m$

$C = \dfrac{8.854 \times 10^{-12} \times (0.3 \times 0.5)}{6 \times 10^{-3}} = 0.221nF$

Then the value of the capacitor consisting of two plates separated by air is calculated as 0.221nF, or 221pF.

The Dielectric

The type of dielectric material being used has an impact on the device's overall capacitance in addition to the conductive plates' overall size and their spacing from one another. In other words, the dielectric's "Permittivity" (ε). The dielectric material used in a capacitor is always an insulator, but the conductive plates of a capacitor are typically made of a metal foil or a metal film, allowing for the flow of electrons and charge. The ability of the various insulating materials used as a capacitor's dielectric to obstruct or pass an electrical charge varies.

The most popular types of insulating materials used to create this dielectric material include: air, paper, polyester, polypropylene, Mylar, ceramic, glass, oil, and a wide range of other materials. The Dielectric Constant, k, is the factor by which the dielectric material, or insulator, increases the capacitance of the capacitor when compared to air, and a dielectric material with a high dielectric constant is a better insulator than a dielectric material with a low dielectric constant.

Since it is relative to free space, the dielectric constant has no dimensions. The actual permittivity, or "complex permittivity," of the dielectric material between the plates is then the product of the permittivity of free space (ε_o) and the relative permittivity (ε_r) of the dielectric material, and is given as:

Complex Permittivity

$$\varepsilon = \varepsilon_o \times \varepsilon_r$$

In other words, if we take the permittivity of free space, ε_o as our base level and make it equal to one, when the vacuum of free space is replaced by some other type of insulating material, their permittivity of its dielectric is referenced to the base dielectric of free space giving a multiplication factor known as "relative permittivity", ε_r. So the value of the complex permittivity, ε will always be equal to the relative permittivity times one.

Typical units of dielectric permittivity, ε or dielectric constant for common materials are: Pure Vacuum = 1.0000, Air = 1.0006, Paper = 2.5 to 3.5, Glass = 3 to 10, Mica = 5 to 7, Wood = 3 to 8 and Metal Oxide Powders = 6 to 20 etc. This then gives us a final equation for the capacitance of a capacitor as:

$$\text{Capacitance, } C = \frac{\varepsilon_0 \varepsilon_r A}{d} \text{ Farads}$$

By "interleaving" more plates within a single capacitor body, it is possible to increase a capacitor's overall capacitance while minimizing its size. A capacitor can have many separate plates connected together in place of just one set of parallel plates, increasing the surface area, A, of the plates.

The two plates of a typical parallel plate capacitor, as depicted above, are designated A and B. As a result, since there are two capacitor plates, we can state that n = 2, where n stands for the number of plates. Therefore, the ultimate solution to our previous equation for a single parallel plate capacitor is:

Capacitance, $C = \dfrac{\varepsilon_0 \varepsilon_r (n-1)A}{d}$ Farads

However, even if the capacitor has two parallel plates, only one of them will be in contact with the dielectric since the other side of each plate will serve as the capacitor's exterior.

We effectively have "one" whole plate in contact with the dielectric if we take the two plates and join them together.

As for a single parallel plate capacitor, $n - 1 = 2 - 1$ which equals 1 as $C = (\varepsilon_0 * \varepsilon_r \times 1 \times A)/d$ is exactly the same as saying: $C = (\varepsilon_0 * \varepsilon_r * A)/d$ which is the standard equation above.

Let's consider a capacitor made up of 9 interleaved plates, then $n = 9$ as shown.

Multi-plate Capacitor

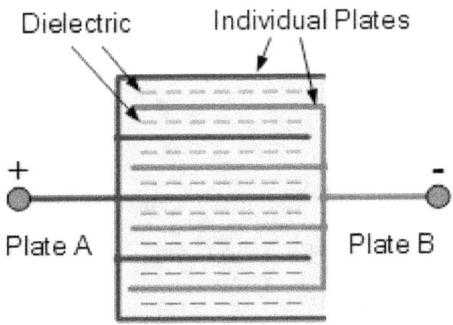

8 mini capacitors in one

15

Five plates are currently attached to lead (A), and four plates are attached to lead (B) (B). The four plates connected to lead B are then in contact with the dielectric on BOTH sides, as opposed to the outer plates connected to lead A, which are only in contact with the dielectric on one side. As stated above, each set of plates has a useful surface area of only eight, and as a result, its capacitance is as follows:

$$C = \frac{\varepsilon_0 \varepsilon_r (n-1) A}{d} = \frac{\varepsilon_0 \varepsilon_r (9-1) A}{d} = \frac{\varepsilon_0 \varepsilon_r 8A}{d}$$

Modern capacitors can be classified according to the characteristics and properties of their insulating dielectric:

- Low Loss, High Stability such as Mica, Low-K Ceramic, Polystyrene.
- Medium Loss, Medium Stability such as Paper, Plastic Film, High-K Ceramic.
- Polarized Capacitors such as Electrolytic's, Tantalum's.

Voltage Rating

All capacitors have a maximum voltage rating and when selecting a capacitor consideration must be given to the amount of voltage to be applied across the capacitor. The maximum amount of voltage that can be applied to the capacitor without damage to its dielectric material is generally given in the data sheets as: WV, (working voltage) or as WV DC, (DC working voltage).

If the voltage applied across the capacitor becomes too great, the dielectric will break down (known as electrical breakdown) and arcing will occur between the capacitor plates resulting in a short-circuit. The working voltage of the capacitor depends on the type of dielectric material being used and its thickness.

The DC working voltage of a capacitor is just that, the maximum DC voltage and NOT the maximum AC voltage as a capacitor with a DC voltage rating of 100 volts DC cannot be safely subjected to an alternating voltage of 100 volts. Since an alternating voltage that has an RMS value of 100 volts will have a peak value of over 141 volts! ($\sqrt{2}$ x 100).

Then a capacitor which is required to operate at 100 volts AC should have a working voltage of at least 200 volts. In practice, a capacitor should be selected so that its working voltage either DC or AC should be at least 50 percent greater than the highest effective voltage to be applied to it.

Another factor which affects the operation of a capacitor is Dielectric Leakage. Dielectric leakage occurs in a capacitor as the result of an unwanted leakage current which flows through the dielectric material.

Generally, it is assumed that the resistance of the dielectric is extremely high and a good insulator blocking the flow of DC current through the capacitor (as in a perfect capacitor) from one plate to the other.

However, if the dielectric material becomes damaged due excessive voltage or over temperature, the leakage current through the dielectric will become extremely high resulting in a rapid loss of charge on the plates and an overheating of the capacitor eventually resulting in premature failure of the capacitor. Then never use a capacitor in a circuit with higher voltages than the capacitor is rated for otherwise it may become hot and explode.

We know that a capacitor's main function is to store electrical charge on its plates. The Capacitance value of a capacitor, which refers to how much electrical charge it can hold on its plates, is determined by three key elements.

- Surface Area – the surface area, A of the two conductive plates which make up the capacitor, the larger the area the greater the capacitance.
- Distance – the distance, d between the two plates, the smaller the distance the greater the capacitance.
- Dielectric Material – the type of material which separates the two plates called the "dielectric", the

higher the permittivity of the dielectric the greater the capacitance.

We also know that a capacitor is composed of metal plates that are not in contact with one another but rather are spaced apart by a substance known as a dielectric. Although a capacitor's dielectric can be air or even a vacuum, it is typically an insulating material that is not conductive, such as waxed paper, glass, mica, various types of plastics, etc. The benefits of the dielectric include the following:

- The dielectric constant is the property of the dielectric material and varies from one material to another increasing the capacitance by a factor of k.
- The dielectric provides mechanical support between the two plates allowing the plates to be closer together without touching.
- Permittivity of the dielectric increases the capacitance.
- The dielectric increases the maximum operating voltage compared to air.

In a wide range of circuits and applications, capacitors can be used to pass audio signals, pulses of alternating current, or other time-varying waveforms while blocking DC current.

Capacitors can be used to smooth the output voltages of power supplies and to remove unwanted spikes from signals that would otherwise tend to cause damage or false triggering of semiconductors or digital components due to their ability to block DC currents. Additionally, capacitors can be used to couple separate amplifier stages that need to be shielded from DC current transmission together or to modify the frequency response of an audio circuit.

A capacitor has infinite impedance (open circuit) when connected to DC supplies, and zero impedance at very high frequencies (short-circuit). It is advisable to choose a capacitor with a voltage rating at least 50% higher than the supply voltage because all capacitors have a maximum working DC voltage rating (WVDC).

CHAPTER-2

TYPES OF CAPACITORS

There are a wide range of different capacitor types available on the market, and each one has a unique set of properties and uses. The variety of capacitor types available includes everything from tiny, delicate trimming capacitors for oscillators or radio circuits to powerful metal-can capacitors for high voltage power smoothing and correction.

The dielectric material used between the plates is typically used to compare the various types of capacitors. For use in radio or "frequency tuning" type circuits, there are variable types of capacitors that work similarly to variable resistors in that we can change the capacitance value. Metal foil is used in commercial capacitor construction, and thin sheets of Mylar or paraffin-impregnated paper are used as the dielectric.

Some capacitors have the appearance of tubes because the insulating dielectric material is sandwiched between the metal foil plates that have been rolled up into a cylinder to create the small package. Small capacitors are frequently made of ceramic materials and sealed by dipping them in epoxy resin. In any case, capacitors are crucial components of electronic circuits, so the following list includes some of the more prevalent varieties.

Dielectric Capacitor

Variable dielectric capacitors are commonly used for tuning transmitters, receivers, and transistor radios, which require a continuous variation in capacitance. Variable dielectric capacitors are multi-plate air-spaced types with fixed plates (stator vanes) and movable plates (rotor vanes) that move in between the fixed plates.

The overall capacitance value is determined by the position of the moving plates in relation to the fixed plates. When the two sets of plates are fully meshed together, the capacitance is generally at its maximum. High voltage tuning capacitors have relatively large spacings or air-gaps between the plates and breakdown voltages in the thousands of volts range.

Variable Capacitor Symbol

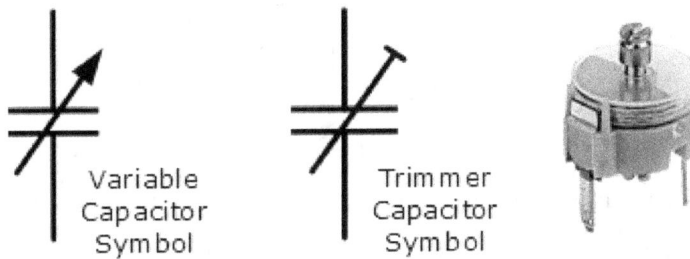

Variable
Capacitor
Symbol

Trimmer
Capacitor
Symbol

Trimmers are preset type variable capacitors that are available in addition to continuously variable capacitors. These are generally small devices that, with the aid of a small screwdriver, can be adjusted or "pre-set" to a specific capacitance value and are available in very small capacitances of 500pF or less and are non-polarized.

Film Capacitor Type

Film capacitors are the most widely available type of capacitor, consisting of a relatively large family of capacitors that differ only in their dielectric properties. Polyester (Mylar), polystyrene, polypropylene, polycarbonate, metalized paper, Teflon, and other materials are examples. Film capacitors are available in capacitance ranges ranging from 5pF to 100uF, depending on the type of capacitor and its voltage rating. Film capacitors are also available in a variety of shapes and case styles, such as:

- Wrap & Fill (Oval & Round) – where the capacitor is wrapped in a tight plastic tape and have the ends filled with epoxy to seal them.
- Epoxy Case (Rectangular & Round) – where the capacitor is encased in a molded plastic shell which is then filled with epoxy.
- Metal Hermetically Sealed (Rectangular & Round) – where the capacitor is encased in a metal tube or can and again sealed with epoxy.

with all the above case styles available in both Axial and Radial Leads.

Film Capacitors which use polystyrene, polycarbonate or Teflon as their dielectrics are sometimes called "Plastic capacitors". The construction of plastic film capacitors is similar to that for paper film capacitors but use a plastic film instead of paper.

The main advantage of plastic film types of capacitor compared to impregnated-paper types is that they operate well under conditions of high temperature, have smaller tolerances, a very long service life and high reliability. Examples of film capacitors are the rectangular metalized film and cylindrical film & foil types as shown below.

Radial Lead Type

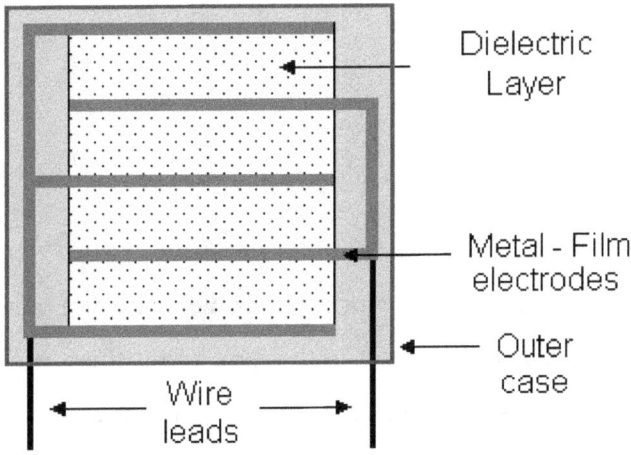

Axial Lead Type

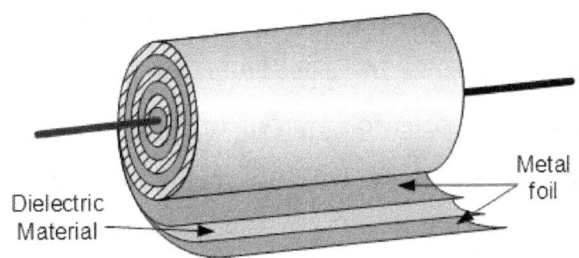

The film and foil types of capacitor are made from long thin strips of thin metal foil with the dielectric material sandwiched together which are wound into a tight roll and then sealed in paper or metal tubes.

Film Capacitor

These film types require a much thicker dielectric film to reduce the risk of tears or punctures in the film, and is therefore more suited to lower capacitance values and larger case sizes.

Metalized foil capacitors have the conductive film metalized sprayed directly onto each side of the dielectric which gives the capacitor self-healing properties and can therefore use much thinner dielectric films. This allows for higher capacitance values and smaller case sizes for a given capacitance. Film and foil capacitors are generally used for higher power and more precise applications.

Ceramic Type Capacitor

The two sides of a small porcelain or ceramic disc are coated with silver to build ceramic capacitors, also known as disc capacitors, which are then stacked together to form capacitors. A single ceramic disc measuring approximately 3-6 mm is used for very low capacitance values. Ceramic capacitors are able to achieve relatively high capacitances in a small physical size because they have a high dielectric constant (High-K).

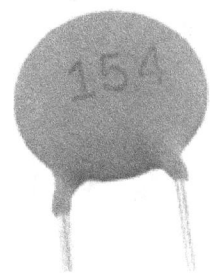

Ceramic Capacitor

They exhibit large non-linear changes in capacitance against temperature and as a result are used as de-coupling or by-pass capacitors as they are also non-polarized devices. Ceramic capacitors have values ranging from a few picofarads to one or two microfarads, (µF) but their voltage ratings are generally quite low.

Ceramic types of capacitors generally have a 3-digit code printed onto their body to identify their capacitance value in pico-farads. Generally, the first two digits indicate the capacitors value and the third digit indicates the number of zero's to be added. For example, a ceramic disc capacitor with the markings 103 would indicate 10 and 3 zero's in pico-farads which is equivalent to 10,000 pF or 10nF.

Likewise, the digits 104 would indicate 10 and 4 zero's in pico-farads which is equivalent to 100,000 pF or 100nF and so on. So on the image of the ceramic capacitor above the numbers 154 indicate 15 and 4 zero's in pico-farads which is equivalent to 150,000 pF or 150nF or 0.15µF. Letter codes are sometimes used to indicate their tolerance value such as: J = 5%, K = 10% or M = 20% etc.

Electrolytic Type Capacitor

Electrolytic Capacitors are generally used when very large capacitance values are required. Here instead of using a very thin metallic film layer for one of the electrodes, a semi-liquid electrolyte solution in the form of a jelly or paste is used which serves as the second electrode (usually the cathode).

The dielectric is a very thin layer of oxide which is grown electro-chemically in production with the thickness of the film being less than ten microns. This insulating layer is so thin that it is possible to make capacitors with a large value of capacitance for a small physical size as the distance between the plates, d is very small.

Electrolytic Capacitor

The majority of electrolytic types of capacitors are Polarized, that is the DC voltage applied to the capacitor terminals must be of the correct polarity, i.e., positive to the positive terminal and negative to the negative terminal as an incorrect polarization will break down the insulating oxide layer and permanent damage may result.

All polarized electrolytic capacitors have their polarity clearly marked with a negative sign to indicate the negative terminal and this polarity must be followed.

Electrolytic Capacitors are generally used in DC power supply circuits due to their large capacitance and small size to help reduce the ripple voltage or for coupling and decoupling applications. One main disadvantage of electrolytic capacitors is their relatively low voltage rating and due to the polarization of electrolytic capacitors, it follows then that they must not be used on AC supplies. Electrolytic's generally come in two basic forms; Aluminium Electrolytic Capacitors and Tantalum Electrolytic Capacitors.

Electrolytic Capacitor

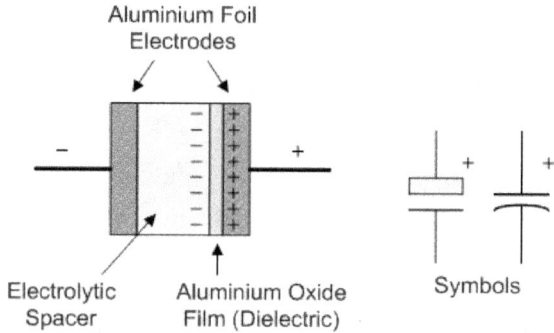

1. Aluminium Electrolytic Type Capacitor

There are basically two types of Aluminium Electrolytic Capacitor, the plain foil type and the etched foil type. The thickness of the aluminium oxide film and high breakdown voltage give these capacitors very high capacitance values for their size.

The foil plates of the capacitor are anodized with a DC current. This anodizing process sets up the polarity of the plate material and determines which side of the plate is positive and which side is negative.

The etched foil type differs from the plain foil type in that the aluminium oxide on the anode and cathode foils has been chemically etched to increase its surface area and permittivity. This gives a smaller sized capacitor than a plain foil type of equivalent value but has the disadvantage of not being able to withstand high DC currents compared to the plain type. Also their tolerance range is quite large at up to 20%. Typical values of capacitance for an aluminium electrolytic capacitor range from 1uF up to 47,000uF.

Etched foil electrolytic's are best used in coupling, DC blocking and by-pass circuits while plain foil types are better suited as smoothing capacitors in power supplies. But aluminium electrolytic's are "polarized" devices so reversing the applied voltage on the leads will cause the insulating layer within the capacitor to become destroyed along with the capacitor. However, the electrolyte used within the capacitor helps heal a damaged plate if the damage is small.

Since the electrolyte has the properties to self-heal a damaged plate, it also has the ability to re-anodize the foil plate. As the anodizing process can be reversed, the electrolyte has the ability to remove the oxide coating from the foil as would happen if the capacitor was connected with a reverse polarity. Since the electrolyte has the ability to conduct electricity, if the aluminium oxide layer was removed or destroyed, the capacitor would allow current to pass from one plate to the other destroying the capacitor, "so be aware".

2. Tantalum Electrolytic Type Capacitor

Tantalum Electrolytic Capacitors and Tantalum Beads, are available in both wet (foil) and dry (solid) electrolytic types with the dry or solid tantalum being the most common. Solid tantalum capacitors use manganese dioxide as their second terminal and are physically smaller than the equivalent aluminium capacitors.

The dielectric properties of tantalum oxide is also much better than those of aluminium oxide giving a lower leakage currents and better capacitance stability which makes them suitable for use in blocking, by-passing, decoupling, filtering and timing applications.

Also, Tantalum Capacitors although polarized, can tolerate being connected to a reverse voltage much more easily than the aluminium types but are rated at much lower working voltages.

Solid tantalum capacitors are usually used in circuits where the AC voltage is small compared to the DC voltage.

However, some tantalum capacitor types contain two capacitors in-one, connected negative-to-negative to form a "non-polarized" capacitor for use in low voltage AC circuits as a non-polarized device. Generally, the positive lead is identified on the capacitor body by a polarity mark, with the body of a tantalum bead capacitor being an oval geometrical shape. Typical values of capacitance range from 47nF to 470uF.

Aluminium & Tantalum Electrolytic Capacitor

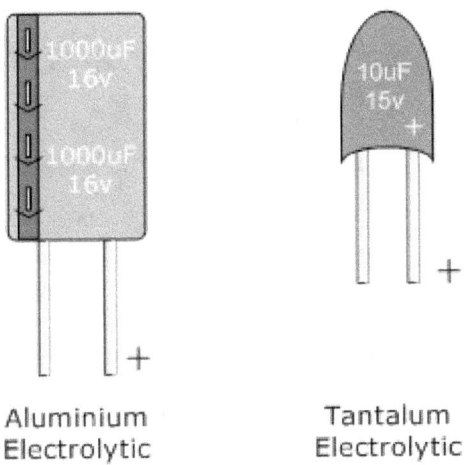

Aluminium
Electrolytic

Tantalum
Electrolytic

Electrolytic's are widely used capacitors due to their low cost and small size but there are three easy ways to destroy an electrolytic capacitor:

- Over-voltage – excessive voltage will cause current to leak through the dielectric resulting in a short circuit condition.
- Reversed Polarity – reverse voltage will cause self-destruction of the oxide layer and failure.
- Over Temperature – excessive heat dries out the electrolytic and shortens the life of an electrolytic capacitor.

CHAPTER-3

CHARACTERISTICS OF CAPACITOR

The characteristics of a capacitors define its temperature, voltage rating and capacitance range as well as its use in a particular application

There are a bewildering array of capacitor characteristics and specifications associated with the humble capacitor and reading the information printed onto the body of a capacitor can sometimes be difficult to understand especially when colors or numeric codes are used.

Each family or type of capacitor uses its own unique set of capacitor characteristics and identification system with some systems being easy to understand, and others that use misleading letters, colors or symbols.

The best way to figure out which capacitor characteristics the label means is to first figure out what type of family the capacitor belongs to whether it is ceramic, film, plastic or electrolytic and from that it may be easier to identify the particular capacitor characteristics.

Even though two capacitors may have exactly the same capacitance value, they may have different voltage ratings. If a smaller rated voltage capacitor is substituted in place of a higher rated voltage capacitor, the increased voltage may damage the smaller capacitor.

We know that with a polarized electrolytic capacitor, the positive lead must go to the positive connection and the negative lead to the negative connection otherwise it may again become damaged. So, it is always better to substitute an old or damaged capacitor with the same type as the specified one. An example of capacitor markings is given below.

Characteristics of Capacitor

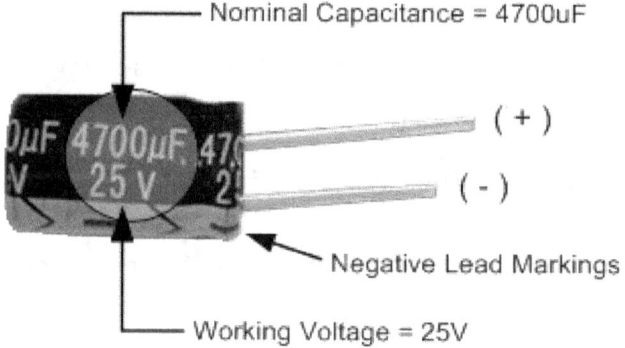

Nominal Capacitance = 4700uF

(+)

(-)

Negative Lead Markings

Working Voltage = 25V

The capacitor, as with any other electronic component, comes defined by a series of characteristics. These Capacitor Characteristics can always be found in the data sheets that the capacitor manufacturer provides to us so here are just a few of the more important ones.

Nominal Capacitance, (C)

The nominal value of the Capacitance, C of a capacitor is the most important of all capacitor characteristics. This value measured in pico-Farads (pF), nano-Farads (nF) or micro-Farads (μF) and is marked onto the body of the capacitor as numbers, letters or colored bands.

The capacitance of a capacitor can change value with the circuit frequency (Hz) y with the ambient temperature. Smaller ceramic capacitors can have a nominal value as low as one pico-

Farad, (1pF) while larger electrolytic's can have a nominal capacitance value of up to one Farad, (1F).

All capacitors have a tolerance rating that can range from -20% to as high as +80% for aluminium electrolytic's affecting its actual or real value. The choice of capacitance is determined by the circuit configuration but the value read on the side of a capacitor may not necessarily be its actual value.

Working Voltage, (WV)

The Working Voltage is another important capacitor characteristic that defines the maximum continuous voltage either DC or AC that can be applied to the capacitor without failure during its working life. Generally, the working voltage printed onto the side of a capacitors body refers to its DC working voltage, (WVDC).

DC and AC voltage values are usually not the same for a capacitor as the AC voltage value refers to the r.m.s. value and NOT the maximum or peak value which is 1.414 times greater. Also, the specified DC working voltage is valid within a certain temperature range, normally -30°C to +70°C.

Any DC voltage in excess of its working voltage or an excessive AC ripple current may cause failure. It follows therefore, that a capacitor will have a longer working life if operated in a cool environment and within its rated voltage. Common working DC voltages are 10V, 16V, 25V, 35V, 50V, 63V, 100V, 160V, 250V, 400V and 1000V and are printed onto the body of the capacitor.

Capacitor Characteristics – Tolerance, (±%)

As with resistors, capacitors also have a Tolerance rating expressed as a plus-or-minus value either in picofarad's (±pF) for low value capacitors generally less than 100pF or as a percentage (±%) for higher value capacitors generally higher than 100pF.

The tolerance value is the extent to which the actual capacitance is allowed to vary from its nominal value and can range anywhere from -20% to +80%. Thus, a 100μF capacitor with a ±20% tolerance could legitimately vary from 80μF to 120μF and still remain within tolerance.

Capacitors are rated according to how near to their actual values they are compared to the rated nominal capacitance with colored bands or letters used to indicated their actual tolerance. The most common tolerance variation for capacitors is 5% or 10% but some plastic capacitors are rated as low as ±1%.

Leakage Current

The dielectric used inside the capacitor to separate the conductive plates is not a perfect insulator resulting in a very small current flowing or "leaking" through the dielectric due to the influence of the powerful electric fields built up by the charge on the plates when applied to a constant supply voltage.

This small DC current flow in the region of nano-amps (nA) is called the capacitors Leakage Current. Leakage current is a result of electrons physically making their way through the dielectric medium, around its edges or across its leads and which will over time fully discharging the capacitor if the supply voltage is removed.

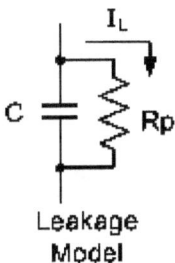

Leakage
Model

When the leakage is very low such as in film or foil type capacitors it is generally referred to as "insulation resistance" (R_p) and can be expressed as a high value resistance in parallel with the capacitor as shown. When the leakage current is high as in electrolytic's it is referred to as a "leakage current" as electrons flow directly through the electrolyte.

Capacitor leakage current is an important parameter in amplifier coupling circuits or in power supply circuits, with the best choices for coupling and/or storage applications being Teflon and the other plastic capacitor types (polypropylene, polystyrene, etc) because the lower the dielectric constant, the higher the insulation resistance.

Electrolytic-type capacitors (tantalum and aluminium) on the other hand may have very high capacitances, but they also have very high leakage currents (typically of the order of about 5-20 µA per µF) due to their poor isolation resistance, and are therefore not suited for storage or coupling applications. Also, the flow of leakage current for aluminium electrolytic's increases with temperature.

Working Temperature, (T)

Changes in temperature around the capacitor affect the value of the capacitance because of changes in the dielectric properties. If the air or surrounding temperature becomes too hot or too cold the capacitance value of the capacitor may change so much as to affect the correct operation of the circuit. The normal working range for most capacitors is -30°C to +125°C with nominal voltage ratings given for a Working Temperature of no more than +70°C especially for the plastic capacitor types.

Generally for electrolytic capacitors and especially aluminium electrolytic capacitor, at high temperatures (over +85°C the liquids within the electrolyte can be lost to evaporation, and the body of the capacitor (especially the small sizes) may become deformed due to the internal pressure and leak outright. Also, electrolytic capacitors cannot be used at low temperatures, below about -10°C, as the electrolyte jelly freezes.

Temperature Coefficient, (TC)

The Temperature Coefficient of a capacitor is the maximum change in its capacitance over a specified temperature range. The temperature coefficient of a capacitor is generally expressed linearly as parts per million per degree centigrade (PPM/°C), or as a percent change over a particular range of temperatures. Some capacitors are nonlinear (Class 2 capacitors) and increase their value as the temperature rises giving them a temperature coefficient that is expressed as a positive "P".

Some capacitors decrease their value as the temperature rises giving them a temperature coefficient that is expressed as a negative "N". For example "P100" is +100 ppm/°C or "N200", which is -200 ppm/°C etc. However, some capacitors do not change their value and remain constant over a certain temperature range, such capacitors have a zero temperature coefficient or "NPO". These types of capacitors such as Mica or Polyester are generally referred to as Class 1 capacitors.

Most capacitors, especially electrolytic's lose their capacitance when they get hot but temperature compensating capacitors are available in the range of at least P1000 through to N5000 (+1000 ppm/°C through to -5000 ppm/°C). It is also possible to connect a capacitor with a positive temperature coefficient in series or parallel with a capacitor having a negative temperature coefficient the net result being that the two opposite effects will cancel each other out over a certain range of temperatures. Another useful application of temperature coefficient capacitors is to use them to cancel out the effect of temperature on other components within a circuit, such as inductors or resistors etc.

Polarization

Capacitor Polarization generally refers to the electrolytic type capacitors but mainly the Aluminium Electrolytic's, with regards to their electrical connection. The majority of electrolytic capacitors are polarized types, that is the voltage connected to the capacitor terminals must have the correct polarity, i.e., positive to positive and negative to negative.

Negative Terminal
Band & Markings

Incorrect polarization can cause the oxide layer inside the capacitor to break down resulting in very large currents flowing through the device resulting in destruction as we have mentioned earlier.

The majority of electrolytic capacitors have their negative, -ve terminal clearly marked with either a black stripe, band, arrows or chevrons down one side of their body as shown, to prevent any incorrect connection to the DC supply.

Some larger electrolytic's have their metal can or body connected to the negative terminal but high voltage types have their metal can insulated with the electrodes being brought out to separate spade or screw terminals for safety.

Also, when using aluminium electrolytic's in power supply smoothing circuits care should be taken to prevent the sum of the peak DC voltage and AC ripple voltage from becoming a "reverse voltage".

Equivalent Series Resistance, (ESR)

The Equivalent Series Resistance or ESR, of a capacitor is the AC impedance of the capacitor when used at high frequencies and includes the resistance of the dielectric material, the DC resistance of the terminal leads, the DC resistance of the connections to the dielectric and the capacitor plate resistance all measured at a particular frequency and temperature.

ESR Model

In some ways, ESR is the opposite of the insulation resistance which is presented as a pure resistance (no capacitive or inductive reactance) in parallel with the capacitor. An ideal capacitor would have only capacitance but ESR is presented as a pure resistance (less than 0.1Ω) in series with the capacitor (hence the name Equivalent Series Resistance), and which is frequency dependent making it a "DYNAMIC" quantity.

As ESR defines the energy losses of the "equivalent" series resistance of a capacitor it must therefore determine the capacitor's overall I^2R heating losses especially when used in power and switching circuits.

Capacitors with a relatively high ESR have less ability to pass current to and from its plates to the external circuit because of their longer charging and discharging RC time constant. The ESR of electrolytic capacitors increases over time as their electrolyte dries out. Capacitors with very low ESR ratings are available and are best suited when using the capacitor as a filter.

As a final note, capacitors with small capacitance's (less than 0.01μF) generally do not pose much danger to humans. However, when their capacitance's start to exceed 0.1μF, touching the capacitor leads can be a shocking experience.

Capacitors have the ability to store an electrical charge in the form of a voltage across themselves even when there is no circuit current flowing, giving them a sort of memory with large electrolytic type reservoir capacitors found in television sets, photo flashes and capacitor banks potentially storing a lethal charge.

As a general rule of thumb, never touch the leads of large value capacitors once the power supply is removed. If you are unsure about their condition or the safe handling of these large capacitors, seek help or expert advice before handling them.

CHAPTER-4:

CAPACITANCE AND CHARGE

Capacitors store electrical energy on their plates in the form of an electrical charge. Capacitance is the measured value of the ability of a capacitor to store an electric charge. This capacitance value also depends on the dielectric constant of the dielectric material used to separate the two parallel plates.

Capacitance is measured in units of the Farad (F), so named after *Michael Faraday*.

Capacitors consist of two parallel conductive plates (usually a metal) which are prevented from touching each other (separated) by an insulating material called the "dielectric". When a voltage is applied to these plates an electrical current flow charging up one plate with a positive charge with respect to the supply voltage and the other plate with an equal and opposite negative charge.

Then, a capacitor has the ability of being able to store an electrical charge Q (units in Coulombs) of electrons. When a capacitor is fully charged there is a potential difference, (p.d.) between its plates, and the larger the area of the plates and/or the smaller the distance between them (known as separation) the greater will be the charge that the capacitor can hold and the greater will be its Capacitance.

Capacitor's ability to store this electrical charge (Q) between its plates is proportional to the applied voltage, V for a capacitor of known capacitance in Farads. Note that capacitance C is ALWAYS positive and never negative.

The greater the applied voltage the greater will be the charge stored on the plates of the capacitor. Likewise, the smaller the applied voltage the smaller the charge. Therefore, the actual charge Q on the plates of the capacitor and can be calculated as:

Charge on a Capacitor

$$Q = C \times V$$

Where:

Q is Charge in Coulombs,

C is Capacitance in Farads,

V is Voltage in Volts

It is sometimes easier to remember this relationship by using pictures. Here the three quantities of Q, C and V have been superimposed into a triangle giving charge at the top with capacitance and voltage at the bottom. This arrangement represents the actual position of each quantity in the *Capacitor Charge* formulas.

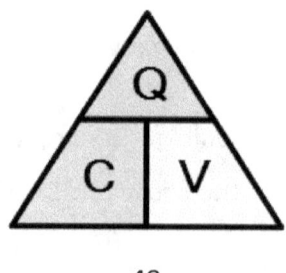

and transposing the above equation gives us the following combinations of the same equation:

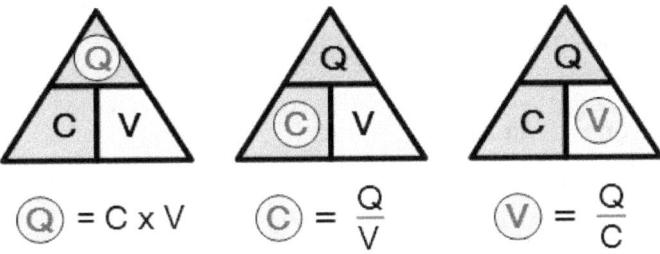

$$\text{\textcircled{Q}} = C \times V \qquad \text{\textcircled{C}} = \frac{Q}{V} \qquad \text{\textcircled{V}} = \frac{Q}{C}$$

Units of: Q measured in Coulombs, V in volts and C in Farads.

Then from above we can define the unit of Capacitance as being a constant of proportionality being equal to the coulomb/volt which is also called a Farad, unit F.

As capacitance represents the capacitor's ability (capacity) to store an electrical charge on its plates we can define one Farad as the *"capacitance of a capacitor which requires a charge of one coulomb to establish a potential difference of one volt between its plates"* as firstly described by Michael Faraday. So, the larger the capacitance, the higher is the amount of charge stored on a capacitor for the same amount of voltage.

The ability of a capacitor to store a charge on its conductive plates gives it its Capacitance value. Capacitance can also be determined from the dimensions or area, A of the plates and the properties of the dielectric material between the plates. A measure of the dielectric material is given by the permittivity, (ε), or the dielectric constant. So, another way of expressing the capacitance of a capacitor is:

Capacitor with Air as its dielectric

$$C = \frac{Q}{V} = \varepsilon \frac{A}{d}$$

Capacitor with a Solid as its dielectric

$$C = \frac{Q}{V} = \varepsilon_0 \varepsilon_r \frac{A}{d}$$

Where A is the area of the plates in square meters, m^2 with the larger the area, the more charge the capacitor can store. d is the distance or separation between the two plates.

The smaller is this distance, the higher is the ability of the plates to store charge, since the -ve charge on the -Q charged plate has a greater effect on the +Q charged plate, resulting in more electrons being repelled off of the +Q charged plate, and thus increasing the overall charge.

ε_0 (epsilon) is the value of the permittivity for air which is 8.854 x 10^{-12} F/m, and ε_r is the permittivity of the dielectric medium used between the two plates.

Parallel Plate Capacitor

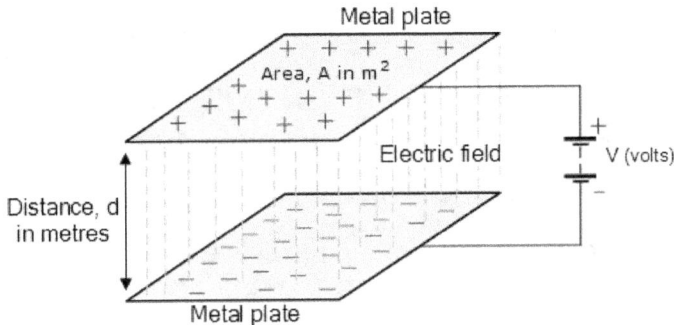

We know that the capacitance of a parallel plate capacitor is proportional to the surface area A and inversely proportional to the distance, d between the two plates and this is true for dielectric medium of air. However, the capacitance value of a capacitor can be increased by inserting a solid medium in between the conductive plates which has a dielectric constant greater than that of air.

Typical values of epsilon ε for various commonly used dielectric materials are: Air = 1.0, Paper = 2.5 − 3.5, Glass = 3 − 10, Mica = 5 − 7 etc.

The factor by which the dielectric material, or insulator, increases the capacitance of the capacitor compared to air is known as the Dielectric Constant, (k). "k" is the ratio of the permittivity of the dielectric medium being used to the permittivity of free space otherwise known as a vacuum.

Therefore, all the capacitance values are related to the permittivity of vacuum. A dielectric material with a high dielectric constant is a better insulator than a dielectric material with a lower dielectric constant. Dielectric constant is a dimensionless quantity since it is relative to free space.

Capacitance Example No1

A parallel plate capacitor consists of two plates with a total surface area of 100 cm². What will be the capacitance in pico-Farads, (pF) of the capacitor if the plate separation is 0.2 cm, and the dielectric medium used is air.

$$C = \varepsilon \frac{A}{d}, \quad \varepsilon = 8.85 \, pF/m$$

$$A = 100 \, cm^2 = 0.01 m^2, \quad d = 0.2 \, cm = 0.002 m$$

$$\therefore \quad C = 8.85 \times 10^{-12} \times \frac{0.01 \, m^2}{0.002 \, m} = 44 \, pF$$

then the value of the capacitor is 44pF.

Charging & Discharging of a Capacitor

Consider the following circuit.

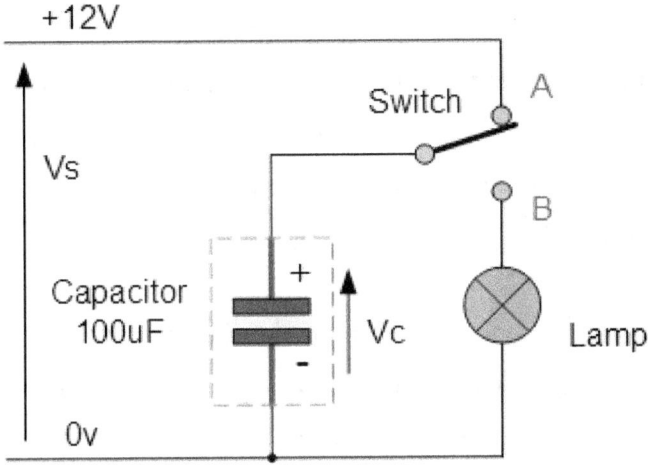

Assume that the capacitor is fully discharged and the switch connected to the capacitor has just been moved to position A. The voltage across the 100uf capacitor is zero at this point and a charging current (i) begins to flow charging up the capacitor exponentially until the voltage across the plates is very nearly equal to the 12v supply voltage. After 5-time constants the current becomes a trickle charge and the capacitor is said to be "fully-charged". Then, $V_C = V_S = 12$ volts.

Once the capacitor is "fully-charged" in theory it will maintain its state of voltage charge even when the supply voltage has been disconnected as they act as a sort of temporary storage device. However, while this may be true of an "ideal" capacitor, a real capacitor will slowly discharge itself over a long period of time due to the internal leakage currents flowing through the dielectric.

This is an important point to remember as large value capacitors connected across high voltage supplies can still maintain a significant amount of charge even when the supply voltage is switched "OFF".

If the switch was disconnected at this point, the capacitor would maintain its charge indefinitely, but due to internal leakage currents flowing across its dielectric the capacitor would very slowly begin to discharge itself as the electrons passed through the dielectric. The time taken for the capacitor to discharge down to 37% of its supply voltage is known as its Time Constant.

If the switch is now moved from position A to position B, the fully charged capacitor would start to discharge through the lamp now connected across it, illuminating the lamp until the capacitor was fully discharged as the element of the lamp has a resistive value.

The brightness of the lamp and the duration of illumination would ultimately depend upon the capacitance value of the capacitor and the resistance of the lamp ($t = R*C$). The larger the value of the capacitor the brighter and longer will be the illumination of the lamp as it could store more charge.

Capacitance Example No2

Calculate the charge in the above capacitor circuit.

$$Q = C \times V$$
$$Q = 100\mu F \times 12v = 1.2 \times 10^{-3} = 1.2mC$$

then the charge on the capacitor is 1.2 millicoulombs.

Current through a Capacitor

Electrical current cannot actually flow through a capacitor as it does a resistor or inductor due to the insulating properties of the dielectric material between the two plates. However, the charging and discharging of the two plates gives the effect that current is flowing.

The current that flows through a capacitor is directly related to the charge on the plates as current is the rate of flow of charge with respect to time. As the capacitors ability to store charge (Q) between its plates is proportional to the applied voltage (V), the relationship between the current and the voltage that is applied to the plates of a capacitor becomes:

Current-Voltage (I-V) Relationship

$$i_{(t)} = C\frac{dv}{dt}$$

As the voltage across the plates increases (or decreases) over time, the current flowing through the capacitance deposits (or removes) charge from its plates with the amount of charge being proportional to the applied voltage. Then both the current and voltage applied to a capacitance are functions of time and are denoted by the symbols, $i_{(t)}$ and $v_{(t)}$.

However, from the above equation we can also see that if the voltage remains constant, the charge will become constant and therefore the current will be zero. In other words, no change in voltage, no movement of charge and no flow of current. This is why a capacitor appears to "block" current flow when connected to a steady state DC voltage.

Capacitance Value – The Farad

We know that the ability of a capacitor to store a charge gives it its capacitance value C, which has the unit of the Farad, F. But the farad is an extremely large unit on its own making it impractical to use, so sub-multiples or fractions of the standard Farad unit are used instead.

To get an idea of how big a Farad really is, the surface area of the plates required to produce a capacitor with a value of just one Farad with a reasonable plate separation of just say 1mm operating in a vacuum. If we rearranging the equation for capacitance above this would give us a plate area of:

$$A = Cd \div 8.85pF/m = (1 \times 0.001) \div 8.85 \times 10^{-12} = 112,994,350 \ m^2$$

or 113 million m^2 which would be equivalent to a plate of more than 10 kilometers x 10 kilometers (over 6 miles) square. That's huge.

Capacitors which have a value of one Farad or more tend to have a solid dielectric and as "One Farad" is such a large unit to use, prefixes are used instead in electronic formulas with capacitor values given in micro-Farads (µF), nano-Farads (nF) and the pico-Farads (pF). For example:

Capacitance Sub-units of the Farad

$$\text{microfarad, } (\mu F) = \frac{1}{1,000,000} F = 1 \times 10^{-6} F$$

$$\text{nanofarad, } (nF) = \frac{1}{1,000,000,000} F = 1 \times 10^{-9} F$$

$$\text{picofarad, } (pF) = \frac{1}{1,000,000,000,000} F = 1 \times 10^{-12} F$$

Convert the following capacitance values:

a) 22nF to µF, b) 0.2µF to nF, c) 550pF to µF

a) 22nF = 0.022µF

b) 0.2µF = 200nF

c) 550pF = 0.00055µF

While one Farad is a large value on its own, capacitors are now commonly available with capacitance values of many hundreds of Farads and have names to reflect this of "Super-capacitors" or "Ultra-capacitors".

These capacitors are electrochemical energy storage devices which utilize a high surface area of their carbon dielectric to deliver much higher energy densities than conventional capacitors and as capacitance is proportional to the surface area of the carbon, the thicker the carbon the more capacitance it has.

Low voltage (from about 3.5V to 5.5V) super-capacitors are capable of storing large amounts of charge due to their high capacitance values as the energy stored in a capacitor is equal to $1/2(C \times V^2)$.

Low voltage super-capacitors are commonly used in portable hand-held devices to replace large, expensive and heavy lithium type batteries as they give battery-like storage and discharge characteristics making them ideal for use as an alternative power source or for memory backup. Super-capacitors used in hand held devices are usually charged using solar cells fitted to the device.

Ultra-capacitor is being developed for use in hybrid electric cars and alternative energy applications to replace large conventional batteries as well as DC smoothing applications in vehicle audio and video systems. Ultra-capacitors can be recharged quickly and have very high energy storage densities making them ideal for use in electric vehicle applications.

Energy in a Capacitor

When a capacitor charges up from the power supply connected to it, an electrostatic field is established which stores energy in the capacitor. The amount of energy in Joules that is stored in this electrostatic field is equal to the energy the voltage supply exerts to maintain the charge on the plates of the capacitor and is given by the formula:

$$\text{Energy, } W = \frac{1}{2}CV^2 \text{ or } \frac{CV^2}{2} \text{ in Joules, (j)}$$

so, the energy stored in the 100uF capacitor circuit above is calculated as:

$$\text{Energy, } W = \frac{CV^2}{2} = \frac{100 \times 10^{-6} \times 12^2}{2} = 7.2\text{mJ}$$

CHAPTER-5

COLOR CODES OF CAPACITOR

Color codes of capacitor are a simple and effective visual way of identifying the capacitance value of a capacitor. There are two common ways to know the capacitive value of a capacitor, by measuring it using a digital multimeter, or by reading the capacitor color codes printed on it. These colored bands represent the capacitance value as per the color code including voltage rating and tolerance.

Sometimes the actual values of capacitance, voltage or tolerance are marked onto the body of a capacitor in the form of alphanumeric characters. However, when the value of the capacitance is of a decimal value problem arise with the marking of the "Decimal Point" as it could easily not be noticed resulting in a misreading of the actual capacitance value.

Instead letters such as p (pico) or n (nano) are used in place of the decimal point to identify its position and the weight of the number. For example, a capacitor can be labelled as, n47 = 0.47nF, 4n7 = 4.7nF or 47n = 47nF and so on.

Also, sometimes capacitors are marked with the capital letter K to signify a value of one thousand pico-Farads, so for example, a capacitor with the markings of 100K would be 100 x 1000pF or 100nF.

To reduce the confusion regarding letters, numbers and decimal points, an international color-coding scheme was developed many years ago as a simple way of identifying capacitor values and tolerances. It consists of colored bands (in spectral order) known commonly as a Capacitor Color Codes system and whose meanings are illustrated below:

Capacitor Color Code Table

Band Color	Digit A	Digit B	Multiplier D	Tolerance (T) > 10pf	Tolerance (T) < 10pf	Temperature Coefficient (TC)
Black	0	0	x1	± 20%	± 2.0pF	
Brown	1	1	x10	± 1%	± 0.1pF	-33×10^{-6}
Red	2	2	x100	± 2%	± 0.25pF	-75×10^{-6}
Orange	3	3	x1,000	± 3%		-150×10^{-6}
Yellow	4	4	x10,000	± 4%		-220×10^{-6}
Green	5	5	x100,000	± 5%	± 0.5pF	-330×10^{-6}
Blue	6	6	x1,000,000			-470×10^{-6}
Violet	7	7				-750×10^{-6}
Grey	8	8	x0.01	+80%,-20%		
White	9	9	x0.1	± 10%	± 1.0pF	

Gold		x0.1	± 5%		
Silver		x0.01	± 10%		

Capacitor Voltage Color Code Table

Band Color	Voltage Rating (V)				
	Type J	Type K	Type L	Type M	Type N
Black	4	100		10	10
Brown	6	200	100	1.6	
Red	10	300	250	4	35
Orange	15	400		40	
Yellow	20	500	400	6.3	6
Green	25	600		16	15
Blue	35	700	630		20
Violet	50	800			
Grey		900		25	25
White	3	1000		2.5	3
Gold		2000			
Silver					

Capacitor Voltage Reference

- Type J – Dipped Tantalum Capacitors.
- Type K – Mica Capacitors.
- Type L – Polyester/Polystyrene Capacitors.
- Type M – Electrolytic 4 Band Capacitors.

- Type N — Electrolytic 3 Band Capacitors.

An example of the use of capacitor color codes is given as:

Metalized Polyester Capacitor

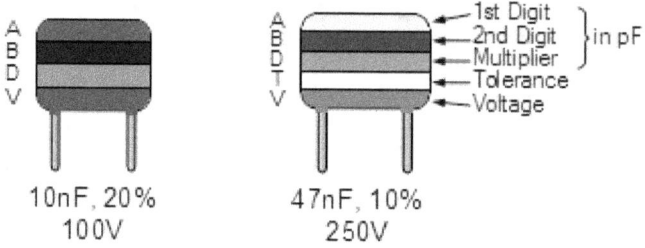

10nF, 20%   47nF, 10%
100V   250V

Disc & Ceramic Capacitor

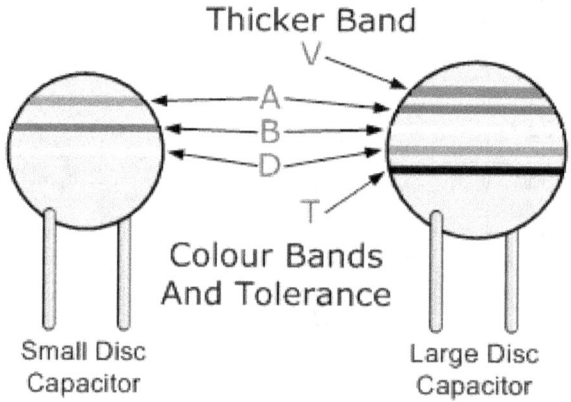

The Capacitor Color Codes system was used for many years on unpolarized polyester and mica molded capacitors. This system of color coding is now obsolete but there are still many "old" capacitors around. Nowadays, small capacitors such as film or disk types conform to the BS1852 Standard and its new replacement, BS EN 60062, where the colors have been replaced by a letter or number coded system.

Generally, the code consists of 2 or 3 numbers and an optional tolerance letter code to identify the tolerance. Where a two number code is used the value of the capacitor only is given in picofarads, for example, 47 = 47 pF and 100 = 100pF etc. A three-letter code consists of the two value digits and a multiplier much like the resistor color codes in the resistors section.

For example, the digits 471 = 47*10 = 470pF. Three-digit codes are often accompanied by an additional tolerance letter code as given below.

Capacitor Tolerance Letter Codes Table

	Letter	B	C	D	F	G	J	K	M	Z
Tolerance	C <10pF ±pF	0.1	0.25	0.5	1	2				
	C >10pF ±%			0.5	1	2	5	10	20	+80 -20

Consider the capacitor below:

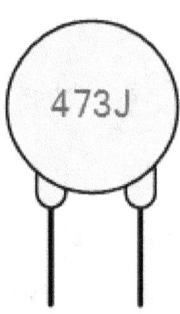

The capacitor on the left is of a ceramic disc type capacitor that has the code 473J printed onto its body. Then the 4 = 1st digit, the 7 = 2nd digit, the 3 is the multiplier in pico-Farads, pF and the letter J is the tolerance and this translates to: 47pF * 1,000 (3 zero's) = 47,000 pF, 47nF or 0.047uF the J indicates a tolerance of +/- 5%

Then by just using numbers and letters as codes on the body of the capacitor we can easily determine the value of its capacitance either in Pico-farad's, Nano-farads or Micro-farads and a list of these "international" codes is given in the following table along with their equivalent capacitances.

Capacitor Letter Codes Table

Picofarad (pF)	Nanofarad (nF)	Microfarad (uF)	Code	Picofarad (pF)	Nanofarad (nF)	Microfarad (uF)	Code
10	0.01	0.00001	100	4700	4.7	0.0047	472
15	0.015	0.000015	150	5000	5.0	0.005	502
22	0.022	0.000022	220	5600	5.6	0.0056	562

33	0.033	0.000033	330	6800	6.8	0.0068	682
47	0.047	0.000047	470	10000	10	0.01	103
100	0.1	0.0001	101	15000	15	0.015	153
120	0.12	0.00012	121	22000	22	0.022	223
130	0.13	0.00013	131	33000	33	0.033	333
150	0.15	0.00015	151	47000	47	0.047	473
180	0.18	0.00018	181	68000	68	0.068	683
220	0.22	0.00022	221	100000	100	0.1	104
330	0.33	0.00033	331	150000	150	0.15	154
470	0.47	0.00047	471	200000	200	0.2	254
560	0.56	0.00056	561	220000	220	0.22	224
680	0.68	0.00068	681	330000	330	0.33	334
750	0.75	0.00075	751	470000	470	0.47	474
820	0.82	0.00082	821	680000	680	0.68	684
1000	1.0	0.001	102	1000000	1000	1.0	105
1500	1.5	0.0015	152	1500000	1500	1.5	155
2000	2.0	0.002	202	2000000	2000	2.0	205
2200	2.2	0.0022	222	2200000	2200	2.2	225
3300	3.3	0.0033	332	3300000	3300	3.3	335

CHAPTER-6

CAPACITORS IN PARALLEL

Capacitors are connected together in parallel when both of its terminals are connected to each terminal of another capacitor. The voltage (Vc) connected across all the capacitors that are connected in parallel is THE SAME. Then, Capacitors in Parallel have a "common voltage" supply across them giving:

$V_{C1} = V_{C2} = V_{C3} = V_{AB} = 12V$

In the following circuit the capacitors, C_1, C_2 and C_3 are all connected together in a parallel branch between points A and B as shown.

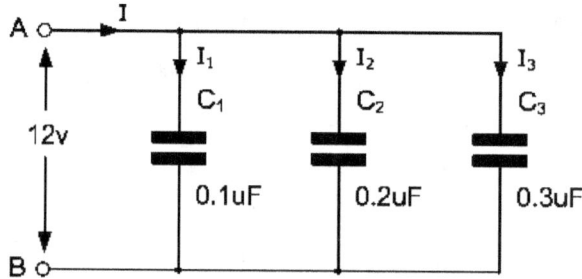

When capacitors are connected together in parallel the total or equivalent capacitance, C_T in the circuit is equal to the sum of all the individual capacitors added together. This is because the top plate of capacitor, C_1 is connected to the top plate of C_2 which is connected to the top plate of C_3 and so on.

The same is also true for the bottom plates of the capacitor. Then it is the same as if the three sets of plates were touching

68

each other and equal to one large single plate thereby increasing the effective plate area in m².

Since capacitance, C is related to plate area ($C = \varepsilon(A/d)$) the capacitance value of the combination will also increase. Then the total capacitance value of the capacitors connected together in parallel is actually calculated by adding the plate area together. In other words, the total capacitance is equal to the sum of all the individual capacitance's in parallel. You may have noticed that the total capacitance of parallel capacitors is found in the same way as the total resistance of series resistors.

We know that the currents flowing through each capacitor are related to the voltage. By applying Kirchhoff's Current Law, (KCL) to the above circuit, we have

$$i_1 = C_1\frac{dv}{dt}, \quad i_2 = C_2\frac{dv}{dt}, \quad i_3 = C_3\frac{dv}{dt}$$

$$i_T = i_1 + i_2 + i_3$$

$$\therefore i_T = C_1\frac{dv}{dt} + C_2\frac{dv}{dt} + C_3\frac{dv}{dt}$$

and this can be re-written as:

$$i_T = \left(C_1 + C_2 + C_3 \right)\frac{dv}{dt}$$

or $\quad i_T = C_T \dfrac{dv}{dt}$

We can define the total or equivalent circuit capacitance, C_T as being the sum of all the individual capacitances add together giving us the generalized equation of:

Parallel Capacitors Equation

$$C_T = C_1 + C_2 + C_3 + ...etc$$

When adding together capacitors in parallel, they must all be converted to the same capacitance units, whether it is µF, nF or pF. Also, we can see that the current flowing through the total capacitance value, C_T is the same as the total circuit current, i_T

We can also define the total capacitance of the parallel circuit from the total stored coulomb charge using the Q = CV equation for charge on a capacitor's plates. The total charge Q_T stored on all the plates equals the sum of the individual stored charges on each capacitor therefore,

$Q_T = Q_1 + Q_2 + Q_3 \quad$ but, $Q = CV$

$\therefore Q_T = CV_T = CV_1 + CV_2 + CV_3$

or $C_T = C_1 + C_2 + C_3$

As the voltage, (V) is common for parallel connected capacitors, we can divide both sides of the above equation through by the voltage leaving just the capacitance and by simply adding together the value of the individual capacitances gives the total capacitance, C_T. Also, this equation is not dependent upon the number of Capacitors in Parallel in the branch, and can therefore be generalized for any number of N parallel capacitors connected together.

Example No1

So, by taking the values of the three capacitors from the above example, we can calculate the total equivalent circuit capacitance C_T as being:

$C_T = C_1 + C_2 + C_3 = 0.1uF + 0.2uF + 0.3uF = 0.6uF$

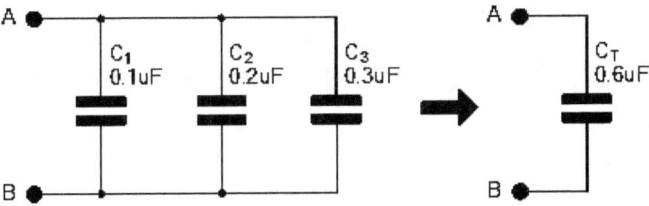

One important point to remember about parallel connected capacitor circuits, the total capacitance (C_T) of any two or more capacitors connected together in parallel will always be GREATER than the value of the largest capacitor in the group as we are adding together values. So, in our example above $C_T = 0.6\mu F$ whereas the largest value capacitor is only $0.3\mu F$.

When 4, 5, 6 or even more capacitors are connected together the total capacitance of the circuit C_T would still be the sum of all the individual capacitors added together and as we know now, the total capacitance of a parallel circuit is always greater than the highest value capacitor.

This is because we have effectively increased the total surface area of the plates. If we do this with two identical capacitors, we have doubled the surface area of the plates which in turn doubles the capacitance of the combination and so on.

Example No2

Determine the combined capacitance in micro-Farads (μF) of the following capacitors when they are connected together in a parallel combination:

- a) two capacitors each with a capacitance of 47nF
- b) one capacitor of 470nF connected in parallel to a capacitor of 1μF

a) Total Capacitance,

$C_T = C_1 + C_2 = 47nF + 47nF = 94nF$ or $0.094\mu F$

b) Total Capacitance,

$C_T = C_1 + C_2 = 470nF + 1\mu F$

therefore, $C_T = 470nF + 1000nF = 1470nF$ or $1.47\mu F$

So, the total or equivalent capacitance, C_T of an electrical circuit containing two or more Capacitors in Parallel is the sum of the all the individual capacitances added together as the effective area of the plates is increased.

CHAPTER-7

CAPACITORS IN SERIES

Capacitors are connected together in series when they are daisy chained together in a single line. With capacitors in series, the charging current (i_C) flowing through the capacitors is THE SAME for all capacitors as it only has one path to follow.

Then, Capacitors in Series all have the same current flowing through them as $i_T = i_1 = i_2 = i_3$ etc. Therefore, each capacitor will store the same amount of electrical charge, Q on its plates regardless of its capacitance. This is because the charge stored by a plate of any one capacitor must have come from the plate of its adjacent capacitor. Therefore, capacitors connected together in series must have the same charge.

$Q_T = Q_1 = Q_2 = Q_3$ …. etc.

Consider the following circuit in which the three capacitors, C_1, C_2 and C_3 are all connected together in a series branch across a supply voltage between points A and B.

Capacitors in a Series Connection

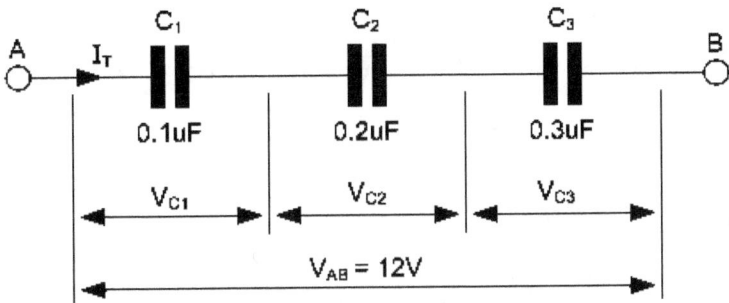

In the previous parallel circuit, we observed that the total capacitance, C_T of the circuit was equal to the sum of all the individual capacitors added together. In a series connected circuit however, the total or equivalent capacitance C_T is calculated differently.

In the series circuit above the right-hand plate of the first capacitor, C_1 is connected to the left-hand plate of the second capacitor, C_2 whose right-hand plate is connected to the left-hand plate of the third capacitor, C_3. Then this series connection means that in a DC connected circuit, capacitor C_2 is effectively isolated from the circuit.

The result of this is that the effective plate area has decreased to the smallest individual capacitance connected in the series chain. Therefore, the voltage drop across each capacitor will be different depending upon the values of the individual capacitances.

Then by applying Kirchhoff's Voltage Law, (KVL) to the above circuit, we get:

$$V_{AB} = V_{C1} + V_{C2} + V_{C3} = 12v$$

$$V_{C1} = \frac{Q_T}{C_1}, \; V_{C2} = \frac{Q_T}{C_2}, \; V_{C3} = \frac{Q_T}{C_3}$$

Since Q = C*V and rearranging for V = Q/C, substituting Q/C for each capacitor voltage V_c in the above KVL equation will give us:

$$V_{AB} = \frac{Q_T}{C_T} = \frac{Q_T}{C_1} + \frac{Q_T}{C_2} + \frac{Q_T}{C_3}$$

dividing each term through by Q gives

Series Capacitors Equation

$$\frac{1}{C_T} = \frac{1}{C_1} + \frac{1}{C_2} + \frac{1}{C_3} + ...etc$$

When adding together Capacitors in Series, the reciprocal (1/C) of the individual capacitors are all added together (just like resistors in parallel) instead of the capacitances themselves. Then the total value for capacitors in series equals the reciprocal of the sum of the reciprocals of the individual capacitances.

Example No1

Taking the three capacitor values from the above example, we can calculate the total equivalent capacitance, C_T for the three capacitors in series as being:

$$\frac{1}{C_T} = \frac{1}{C_1} + \frac{1}{C_2} + \frac{1}{C_3}$$

$$\left(\frac{1}{0.1\,\mu F} + \frac{1}{0.2\,\mu F} + \frac{1}{0.3\,\mu F} \right) = 18.33 \times 10^6$$

$$\frac{1}{C_T} = 18.33 \times 10^6 \quad \therefore\ C_T = \frac{1}{18.33 \times 10^6} = 0.055\,\mu F$$

One important point to remember about capacitors that are connected together in a series configuration. The total circuit capacitance (C_T) of any number of capacitors connected together in series will always be LESS than the value of the smallest capacitor in the series string. In our example above, the total capacitance C_T was calculated as being 0.055µF but the value of the smallest capacitor in the series chain is only 0.1µF.

This reciprocal method of calculation can be used for calculating any number of individual capacitors connected together in a single series network. If, however, there are only two capacitors in series, then a much simpler and quicker formula can be used. This is given as:

$$C_T = \frac{C_1 \times C_2}{C_1 + C_2}$$

If the two series connected capacitors are equal and of the same value, that is: $C_1 = C_2$, we can simplify the above equation further as follows to find the total capacitance of the series combination.

$$C_T = \frac{C^2}{2C} = \frac{C}{2} = \frac{1}{2}C$$

Then we can see that if and only if the two series connected capacitors are the same and equal, then the total capacitance, C_T will be exactly equal to one half of the capacitance value, that is: C/2.

With series connected resistors, the sum of all the voltage drops across the series circuit will be equal to the applied voltage V_S (Kirchhoff's Voltage Law) and this is also true about capacitors in series.

With series connected capacitors, the capacitive reactance of the capacitor acts as an impedance due to the frequency of the supply. This capacitive reactance produces a voltage drop across each capacitor, therefore the series connected capacitors act as a capacitive voltage divider network.

The result is that the voltage divider formula applied to resistors can also be used to find the individual voltages for two capacitors in series. Then:

$$V_{CX} = V_S \frac{C_T}{C_X}$$

Where: C_X is the capacitance of the capacitor in question, V_S is the supply voltage across the series chain and V_{CX} is the voltage drop across the target capacitor.

Example No2

Find the overall capacitance and the individual rms voltage drops across the following sets of two capacitors in series when connected to a 12V AC supply.

- a) two capacitors each with a capacitance of 47nF
- b) one capacitor of 470nF connected in series to a capacitor of 1µF

a) Total Equal Capacitance,

$$C_T = \frac{C_1 \times C_2}{C_1 + C_2} = \frac{47nF \times 47nF}{47nF + 47nF} = 23.5nF$$

Voltage drop across the two identical 47nF capacitors,

$$V_{C1} = \frac{C_T}{C_1} \times V_T = \frac{23.5nF}{47nF} \times 12V = 6 \text{volts}$$

$$V_{C2} = \frac{C_T}{C_2} \times V_T = \frac{23.5nF}{47nF} \times 12V = 6 \text{volts}$$

b) Total Unequal Capacitance,

$$C_T = \frac{C_1 \times C_2}{C_1 + C_2} = \frac{47nF \times 47nF}{47nF + 47nF} = 23.5nF$$

Voltage drop across the two non-identical Capacitors: $C_1 = 470nF$ and $C_2 = 1\mu F$.

$$V_{C1} = \frac{C_T}{C_1} \times V_T = \frac{320nF}{470nF} \times 12 = 8.16 \text{volts}$$

$$V_{C2} = \frac{C_T}{C_2} \times V_T = \frac{320nF}{1uF} \times 12 = 3.84 \text{volts}$$

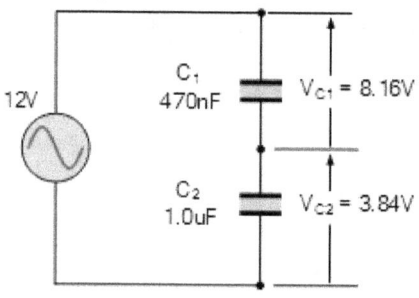

Since Kirchhoff's voltage law applies to this and every series connected circuit, the total sum of the individual voltage drops will be equal in value to the supply voltage, V_S. Then 8.16 + 3.84 = 12V.

Note also that if the capacitor values are the same, 47nF in our first example, the supply voltage will be divided equally across each capacitor as shown. This is because each capacitor in the series chain shares an equal and exact amount of charge ($Q = C \times V = 0.564\mu C$) and therefore has half (or percentage fraction for more than two capacitors) of the applied voltage, V_S.

However, when the series capacitor values are different, the larger value capacitor will charge itself to a lower voltage and the smaller value capacitor to a higher voltage, and in our second example above this was shown to be 3.84 and 8.16 volts respectively. This difference in voltage allows the capacitors to maintain the same amount of charge, Q on the plates of each capacitor as shown.

$$Q_{C1} = V_{C1} \times C_1 = 8.16V \times 470nF = 3.84\mu C$$

$$Q_{C2} = V_{C2} \times C_2 = 3.84V \times 1uF = 3.84\mu C$$

Note that the ratios of the voltage drop across the two capacitors connected in series will always remain the same regardless of the supply frequency as their reactance, X_C will remain proportionally the same.

Then the two voltage drops of 8.16 volts and 3.84 volts above in our simple example will remain the same even if the supply frequency is increased from 100Hz to 100kHz.

Although the voltage drops across each capacitor will be different for different values of capacitance, the coulomb charge across the plates will be equal because the same amount of current flow exists throughout a series circuit as all the capacitors are being supplied with the same number or quantity of electrons.

In other words, if the charge across each capacitors plates is the same, as Q is constant, then as its capacitance decreases the voltage drop across the capacitors plates increases, because the charge is large with respect to the capacitance. Likewise, a larger capacitance will result in a smaller voltage drop across its plates because the charge is small with respect to the capacitance.

CHAPTER-8

CAPACITANCE IN AC CIRCUITS

Capacitors that are connected to a sinusoidal supply produce reactance from the effects of supply frequency and capacitor size

Capacitance in AC Circuits results in a time-dependent current which is shifted in phase by 90° with respect to the supply voltage producing an effect known as capacitive reactance.

When capacitors are connected across a direct current DC supply voltage, their plates charge-up until the voltage value across the capacitor is equal to that of the externally applied voltage. The capacitor will hold this charge indefinitely, acting like a temporary storage device as long as the applied voltage is maintained.

During this charging process, an electric current (i) flows into the capacitor which results in its plates beginning to hold an electrostatic charge. This charging process is not instantaneous or linear as the strength of the charging current is at its maximum when the capacitors plates are uncharged, decreasing exponentially over time until the capacitor is fully-charged.

This is due to the electrostatic field between the plates opposes any changes to the potential difference across the plates at a rate that is equal to the rate of change of the electrical charge on the plates. The property of a capacitor to store a charge on its plates is called its capacitance, (C).

In this way, a capacitors charging current can be defined as: $i = CdV/dt$. Once the capacitor is "fully-charged" the capacitor blocks the flow of any more electrons onto its plates as they have become saturated. However, if we apply an alternating current or AC supply, the capacitor will alternately charge and discharge at a rate determined by the frequency of the supply. Then the Capacitance in AC circuits varies with frequency as the capacitor is being constantly charged and discharged.

We know that the flow of electrons onto the plates of a capacitor is directly proportional to the rate of change of the voltage across those plates. Then, we can see that for capacitance in AC circuits they like to pass current when the voltage across its plates is constantly changing with respect to time such as in AC signals.

However, they do not like to pass current when the applied voltage is of a constant steady state value such as in DC signals. Consider the circuit below.

AC Capacitor Circuit

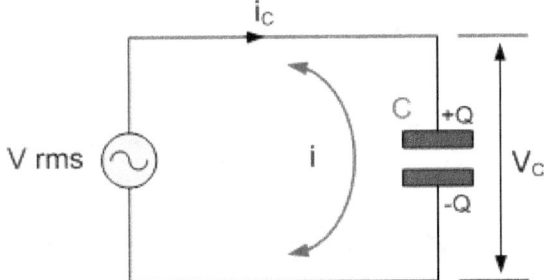

In the purely capacitive circuit above, the capacitor is connected directly across the AC supply voltage. As the supply voltage increases and decreases, the capacitor charges and discharges with respect to this change. We know that the charging current is directly proportional to the rate of change of the voltage across the plates with this rate of change at its greatest as the supply voltage crosses over from its positive half cycle to its negative half cycle or vice versa at points, 0° and 180° along the sine wave.

Consequently, the least voltage rate-of-change occurs when the AC sine wave crosses over at its maximum positive peak ($+V_{MAX}$) and its minimum negative peak, ($-V_{MAX}$). At these two positions within the cycle, the sinusoidal voltage is constant, therefore its rate-of-change is zero, so dv/dt is zero, resulting in zero current change within the capacitor. Thus, when dv/dt = 0, the capacitor acts as an open circuit, so i = 0 and this is shown below.

AC Capacitor Phasor Diagram

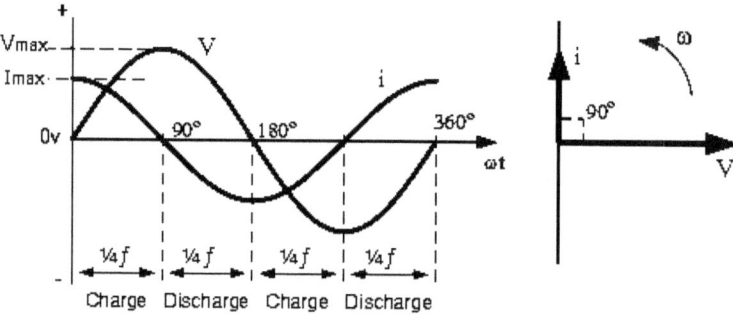

At 0° the rate of change of the supply voltage is increasing in a positive direction resulting in a maximum charging current at that instant in time. As the applied voltage reaches its maximum peak value at 90° for a very brief instant in time the supply voltage is neither increasing or decreasing so there is no current flowing through the circuit.

As the applied voltage begins to decrease to zero at 180°, the slope of the voltage is negative so the capacitor discharges in the negative direction. At the 180° point along the line the rate

86

of change of the voltage is at its maximum again so maximum current flows at that instant and so on.

Then we can say that for capacitors in AC circuits the instantaneous current is at its minimum or zero whenever the applied voltage is at its maximum and likewise the instantaneous value of the current is at its maximum or peak value when the applied voltage is at its minimum or zero.

From the waveform above, we can see that the current is leading the voltage by 1/4 cycle or 90° as shown by the vector diagram. Then we can say that in a purely capacitive circuit the alternating voltage lags the current by 90°.

We know that the current flowing through the capacitance in AC circuits is in opposition to the rate of change of the applied voltage. But just like resistors, capacitors also offer some form of resistance against the flow of current. For capacitors in AC circuits opposition is known as Reactance, and as we are dealing with capacitor circuits, it is therefore known as Capacitive Reactance. Thus, capacitance in AC circuits suffer from Capacitive Reactance.

Capacitance in AC Circuits – Reactance

Capacitive Reactance in a purely capacitive circuit is the opposition to current flow in AC circuits only. Like resistance, reactance is also measured in Ohm's but is given the symbol X to distinguish it from a purely resistive value. As reactance is a quantity that can also be applied to Inductors as well as Capacitors, when used with capacitors it is more commonly known as Capacitive Reactance.

For capacitors in AC circuits, capacitive reactance is given the symbol Xc. Then we can actually say that Capacitive Reactance is a capacitors resistive value that varies with frequency. Also, capacitive reactance depends on the capacitance of the capacitor in Farads as well as the frequency of the AC waveform and the formula used to define capacitive reactance is given as:

Capacitive Reactance

$$X_C = \frac{1}{2\pi f C} = \frac{1}{\omega C}$$

Where: F is in Hertz and C is in Farads. $2\pi f$ can also be expressed collectively as the Greek letter Omega, ω to denote an angular frequency.

From the capacitive reactance formula above, it can be seen that if either of the Frequency or Capacitance where to be increased the overall capacitive reactance would decrease. As the frequency approaches infinity, the capacitors reactance would reduce to zero acting like a perfect conductor.

However, as the frequency approaches zero or DC, the capacitors reactance would increase up to infinity, acting like a very large resistance. This means then that capacitive reactance is "Inversely proportional" to frequency for any given value of Capacitance and this shown below:

Capacitive Reactance against Frequency

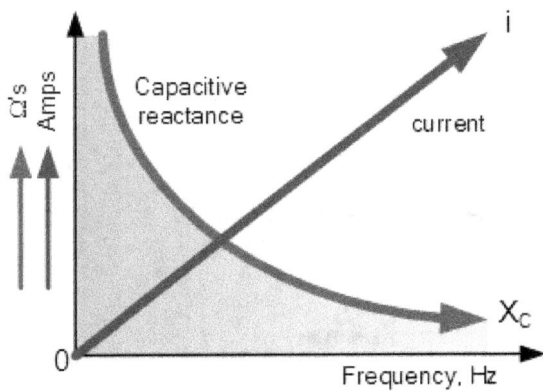

The capacitive reactance of the capacitor decreases as the frequency across it increases therefore capacitive reactance is inversely proportional to frequency.

The opposition to current flow, the electrostatic charge on the plates (its AC capacitance value) remains constant as it becomes easier for the capacitor to fully absorb the change in charge on its plates during each half cycle.

As the frequency increases the current flowing through the capacitor also increases in value due to the rate of voltage change across its plates increases.

We observe that at DC a capacitor has infinite reactance (open-circuit), at very high frequencies a capacitor has zero reactance (short-circuit).

Capacitance in AC Circuits Example No1

Find the rms current flowing in an AC capacitive circuit when a 4µF capacitor is connected across an 880V, 60Hz supply.

$$X_C = \frac{1}{2\pi f C} = \frac{1}{2\pi \times 60 \times (4 \times 10^{-6})} = 663\Omega$$

$$I_{RMS} = \frac{V_{RMS}}{X_C} = \frac{880V}{663\Omega} = 1.33 Amperes$$

In AC circuits, the sinusoidal current through a capacitor, which leads the voltage by 90°, varies with frequency as the capacitor is being constantly charged and discharged by the applied voltage. The AC impedance of a capacitor is known as Reactance and as we are dealing with capacitor circuits, more commonly called Capacitive Reactance, X_C

Capacitance in AC Circuits Example No2.

When a parallel plate capacitor was connected to a 60Hz AC supply, it was found to have a reactance of 390 ohms. Calculate the value of the capacitor in micro-farads.

$$X_C = \frac{1}{2\pi f C} \quad \therefore C = \frac{1}{2\pi f X_C}$$

$$C = \frac{1}{2\pi \times 60 \times 390} = 6.8 uF$$

This capacitive reactance is inversely proportional to frequency and produces the opposition to current flow around a capacitive AC circuit.

CHAPTER-9

CAPACITIVE VOLTAGE DIVIDER

Voltage divider circuits can be built with reactive components just as easily as they can with fixed value resistors. A capacitive voltage divider network, like a resistive circuit, is unaffected by changes in supply frequency, despite the use of capacitors, which are reactive elements, because each capacitor in the series chain is affected equally by changes in supply frequency. However, before we can examine a capacitive voltage divider circuit in greater detail, we must first understand capacitive reactance and how it affects capacitors at different frequencies.

We know that a capacitor consists of two parallel conductive plates separated by an insulator, and has a positive (+) charge on one plate, and an opposite negative (−) charge on the other. We also know that when connected to a DC (direct current) supply, once the capacitor is fully charged, the insulator (called the dielectric) blocks the flow of current through it.

Typical Capacitor

A capacitor opposes current flow just like a resistor, but unlike a resistor which dissipates its unwanted energy in the form of heat, a capacitor stores energy on its plates when it charges and releases or gives back the energy into the connected circuit when it discharges.

This ability of a capacitor to oppose or "react" against current flow by storing charge on its plates is called "reactance", and as this reactance relates to a capacitor it is therefore called Capacitive Reactance (X_c), and like resistance, reactance is also measured in Ohm's.

When a fully discharged capacitor is connected across a DC supply such as a battery or power supply, the reactance of the capacitor is initially extremely low and maximum circuit current flows through the capacitor for a very short period time as the capacitor's plates charge up exponentially.

After a period of time equal to about "5RC" or 5-time constants, the plates of the capacitor are fully charged equaling the supply voltage and no further current flows. At this point the reactance of the capacitor to DC current flow is at its maximum in the mega-ohm region, almost an open-circuit, and this is why capacitors block DC.

Now if we connect the capacitor to an AC (alternating current) supply which is continually reversing polarity, the effect on the capacitor is that its plates are continuously charging and discharging in relationship to the applied alternating supply voltage. This means that a charging and discharging current is always flowing in and out of the capacitors plates, and if we have a current flow we must also have a value of reactance to oppose it. But what value would it be and what factors determine the value of capacitive reactance.

We know that the amount of charge, (Q) present on a capacitors plates is proportional to the applied voltage and capacitance value of the capacitor. As the applied alternating supply voltage, (Vs) is constantly changing in value the charge on the plates must also be changing in value.

If the capacitor has a larger capacitance value, then for a given resistance, R it takes longer to charge the capacitor as $\tau = RC$, which means that the charging current is flowing for a longer period of time. A higher capacitance results in a small value of reactance, Xc for a given frequency.

Similarly, if the capacitor has a small capacitance value, then a shorter RC time constant is required to charge the capacitor which means that the current will flow for a shorter period of time. A smaller capacitance results in a higher value of reactance, Xc. Then we can see that larger currents mean smaller reactance, and smaller currents mean larger reactance. Therefore, capacitive reactance is inversely proportional to the capacitance value of the capacitor, $X_C \propto^{-1} C$.

However, capacitive reactance is determined by more than just capacitance. When the applied alternating current has a low frequency, the reactance has more time to accumulate for a given RC time constant and oppose the current, indicating a high value of reactance.

Similarly, if the applied frequency is high, there is little time for the reactance to build up and oppose the current, resulting in a larger current flow, indicating a smaller reactance.

A capacitor is an impedance, and the magnitude of this impedance varies with frequency. As a result, higher frequencies result in lower reactance, while lower frequencies result in higher reactance. Therefore, Capacitive Reactance, Xc (its complex impedance) is inversely proportional to both capacitance and frequency and the standard equation for capacitive reactance is given as:

Capacitive Reactance Formula

$$Xc = \frac{1}{2\pi f C}$$

- Where:
- Xc = Capacitive Reactance in Ohms, (Ω)
- π (pi) = a numeric constant of 3.142
- f = Frequency in Hertz, (Hz)

- C = Capacitance in Farads, (F)

Voltage Distribution in Series Capacitors

We observed how the opposition to the charging and discharging currents of a capacitor are determined not only by its capacitance value but also by the frequency of the supply, let's look at how this affects two capacitors connected in series forming a capacitive voltage divider circuit.

Capacitive Voltage Divider

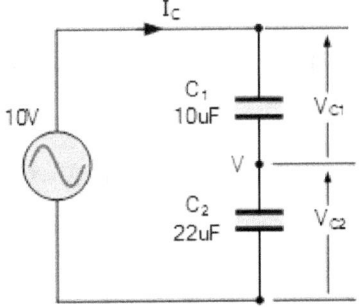

Consider the two capacitors, C1 and C2 connected in series across an alternating supply of 10 volts. As the two capacitors are in series, the charge Q on them is the same, but the voltage across them will be different and related to their capacitance values, as V = Q/C.

Voltage divider circuits may be constructed from reactive components just as easily as they may be constructed from resistors as they both follow the voltage divider rule. Take this capacitive voltage divider circuit, for instance.

The voltage across each capacitor can be calculated in a number of ways. One such way is to find the capacitive reactance value of each capacitor, the total circuit impedance, the circuit current and then use them to calculate the voltage drop, for example:

Capacitive Voltage Divider Example No1

Using the two capacitors of 10uF and 22uF in the series circuit above, calculate the rms voltage drops across each capacitor when subjected to a sinusoidal voltage of 10 volts rms at 80Hz.

Capacitive Reactance of 10uF capacitor

$$X_{C1} = \frac{1}{2\pi f C} = \frac{1}{2\pi \times 80 \times 10 \times 10^{-6}} = 200\Omega$$

Capacitive Reactance of 22uF capacitor

$$X_{C2} = \frac{1}{2\pi f C} = \frac{1}{2\pi \times 80 \times 22 \times 10^{-6}} = 90\Omega$$

Total capacitive reactance of series circuit — Note that reactance's in series are added together just like resistors in series.

$$Xc_{(total)} = X_{C1} + X_{C2}$$
$$= 200\Omega + 90\Omega = 290\Omega$$

or:

$$C_T = \frac{C_1 \times C_2}{C_1 + C_2} = \frac{10uF \times 22uF}{10uF + 22uF} = 6.88uF$$

$$X_C = \frac{1}{2\pi f C_T} = \frac{1}{2\pi \times 80 \times 6.88uF} = 290\Omega$$

Circuit current

$$I = \frac{E}{Xc} = \frac{10V}{290\Omega} = 34.5mA$$

Then the voltage drop across each capacitor in series capacitive voltage divider will be:

$$V_{C1} = I \times X_{C1} = 34.5mA \times 200\Omega = 6.9V$$

$$V_{C2} = I \times X_{C2} = 34.5mA \times 90\Omega = 3.1V$$

When the capacitor values are different, the smaller value capacitor will charge itself to a higher voltage than the larger value capacitor, and in our example above this was 6.9 and 3.1 volts respectively. Since Kirchhoff's voltage law applies to this and every series connected circuit, the total sum of the individual voltage drops will be equal in value to the supply voltage, V_s and 6.9 + 3.1 does indeed equal 10 volts.

Note that the ratios of the voltage drop across the two capacitors connected in a series capacitive voltage divider circuit will always remain the same regardless of the supply frequency. Then the two voltage drops of 6.9 volts and 3.1 volts above in our simple example will remain the same even if the supply frequency is increased from 80Hz to 8000Hz as shown.

Capacitive Voltage Divider Example No2

Using the same two capacitors, calculate the capacitive voltage drop at 8,000Hz (8kHz).

$$X_{C1} = \frac{1}{2\pi f C} = \frac{1}{2\pi \times 8000 \times 10 \text{uF}} = 2\Omega$$

$$X_{C1} = \frac{1}{2\pi f C} = \frac{1}{2\pi \times 8000 \times 22 \text{uF}} = 0.9\Omega$$

$$I = \frac{V}{X_{C(Total)}} = \frac{10}{2.9} = 3.45 \text{ Amps}$$

$$\therefore V_{C1} = I \times X_{C1} = 3.45\text{A} \times 2\Omega = 6.9\text{ V}$$
and
$$V_{C2} = I \times X_{C2} = 3.45\text{A} \times 0.9\Omega = 3.1\text{V}$$

While the voltage ratios across the two capacitors may stay the same, as the supply frequency increases, the combined capacitive reactance decreases, and therefore so too does the total circuit impedance. This reduction in impedance causes more current to flow. For example, at 80Hz we calculated the circuit current above to be about 34.5mA, but at 8kHz, the supply current increased to 3.45A, 100 times more. Therefore, the current flowing through a capacitive voltage divider is proportional to frequency or $I \propto f$.

We have seen here that a capacitor divider is a network of series connected capacitors, each having an AC voltage drop across it. As capacitive voltage dividers use the capacitive reactance value of a capacitor to determine the actual voltage drop, they can only be used on frequency driven supplies and as such do not work as DC voltage dividers. This is mainly due to the fact that capacitors block DC and therefore no current flows.

Colpitts oscillators, capacitive touch-sensitive screens that change their output voltage when touched by a person's finger, capacitive voltage divider circuits used as a less expensive alternative to mains transformers in dropping high voltages such as in mains connected circuits that use low voltage electronics or ICs, etc.

The voltage division across a capacitive voltage divider circuit will always remain the same, maintaining a steady voltage divider, because, as we now know, the reactance of both capacitors changes with frequency (at the same rate).

CHAPTER-10

ULTRACAPACITORS

Ultracapacitors are electrical energy storage devices which can store a significant amount of electrical charge. Unlike resistors, which dissipate energy as heat, ideal ultracapacitors do not lose energy. It has also been observed that the simplest form of a capacitor is two parallel conducting metal plates which are separated by an insulating material, such as air, mica, paper, ceramic, etc., and called the dielectric through a distance, "d". Capacitors store energy due to their ability to store charge, with the amount of charge stored on a capacitor depending on the voltage, V, applied across its plates, and the higher the voltage, the more charge the capacitor will store as: $Q \propto V$.

A Typical Ultracapacitor

A capacitor has a constant of proportionality, called capacitance, symbol C, which represents the capacitor's ability or capacity to store an electrical charge with the amount of charge depending on a capacitor capacitance value as: $Q \propto C$.

Then we can see that there is a relationship between the charge, Q, voltage V and capacitance C, and the larger the capacitance, the higher is the amount of charge stored on a capacitor for the same amount of voltage and we can define this relationship for a capacitor as being:

Charge on a Capacitor

$$Q = C \times V$$

Where: Q (Charge, in Coulombs) = C (Capacitance, in Farads) times V (Voltage, in Volts)

The unit of capacitance is the coulomb/volt, which is also called the Farad (F) [named after M. Faraday] with one farad being defined as the capacitance of a capacitor, which requires a charge of 1 coulomb to establish a potential difference of 1 volt between its two plates.

But a conventional one farad capacitor would be very large for most practical electronic applications, hence much smaller units like the microfarad (μF), nanofarad (nF) and picofarad (pF) are commonly used where:

- Microfarad (μF) $1\mu F = 1/1,000,000 = 0.000001 = 10^{-6}$ F
- Nanofarad (nF) $1nF = 1/1,000,000,000 = 0.000000001 = 10^{-9}$ F
- Picofarad (pF) $1pF = 1/1,000,000,000,000 = 0.000000000001 = 10^{-12}$ F

However, there is another type of capacitor available, called an Ultracapacitor or Supercapacitor which can provide values from a few milli-farads (mF) to tens of farads of capacitance in a very small size allowing for much more electrical energy to be stored between their plates. We know that the energy stored in a capacitor is given by the equation:

$$E = \frac{1}{2}CV^2 \text{ in Joules}$$

Where: E is the energy stored in the electric field in joules, V is the potential difference across the plates and C is the capacitance of the capacitor in farads and defined as:

$$C = \varepsilon\frac{A}{d} \text{ in Farads}$$

Where: ε is the permittivity of the material between the plates, A is the area of the plates, and d is the separation of the plates.

Ultracapacitors are another type of capacitor which is constructed to have a large conductive plate, called an electrode, surface area (A) as well as a very small distance (d) between them. Unlike conventional capacitors that use a solid and dry dielectric material such as Teflon, Polyethylene, Paper, etc., the ultracapacitor uses a liquid or wet electrolyte between its electrodes making it more of an electrochemical device similar to an electrolytic capacitor.

Although an ultracapacitor is a type of electrochemical device, no chemical reactions are involved in the storing of its electrical energy. This means that the ultra-capacitor remains effectively an electrostatic device storing its electrical energy in the form of an electric field between its two conducting electrodes as shown.

An Ultracapacitors Construction

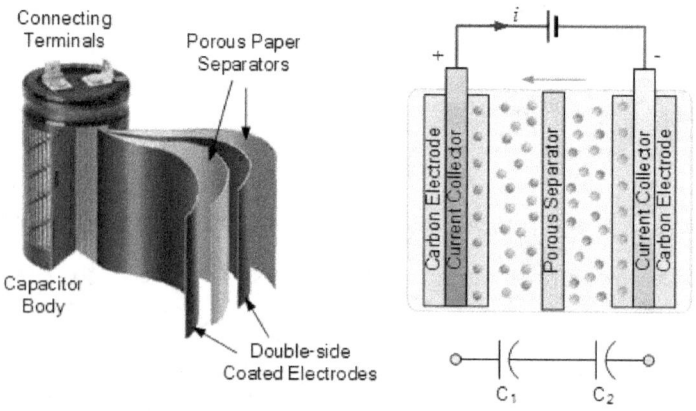

The double-sided coated electrodes are made from graphite carbon in the form of activated conductive carbon, carbon nanotubes or carbon gels. A porous paper membrane called a separator keeps the electrodes apart but allows positive ion to pass through while blocking the larger electrons. Both the paper separator and carbon electrodes are impregnated with the liquid electrolyte with an aluminium foil used in between the two to act as the current collector making electrical connection to the ultracapacitors solder tabs.

The double layer construction of the carbon electrodes and separator may be very thin but their effective surface area into the thousands of meters squared when coiled up together. Then in order to increase the capacitance of an ultra-capacitor, it is obvious that we need to increase the contact surface area, A (in m^2) without increasing the capacitors physical size, or use a special type of electrolyte to increase the available positive ions to increase conductivity.

Then ultra-capacitors make excellent energy storage devices because of their high values of capacitance up into the hundreds of farads, due to the very small distance d or separation of their plates and the electrodes high surface area A for the formation on the surface of a layer of electrolytic ions forming a double layer. This construction effectively creates two capacitors, one at each carbon electrode, giving the ultracapacitor the secondary name of "double layer capacitor" forming two capacitors in series.

However, the problem with this small size is that the voltage across the capacitor can only be very low as the rated voltage of the ultra-capacitor cell is determined mainly by the decomposition voltage of the electrolyte. Then a typical capacitor cell has a working voltage of between 1 to 3 volts, depending on the electrolyte used, which can limit the amount of electrical energy it can store.

In order to store charge at a reasonable voltage ultracapacitors have to be connected in series. Unlike electrolytic and electrostatic capacitors, ultra-capacitors are characterized by their low terminal voltage. In order to increase their rated terminal voltage to tens of volts, ultracapacitor cells must be connected in series, or in parallel to achieve higher capacitance values as shown.

Increasing An Ultracapacitors Value

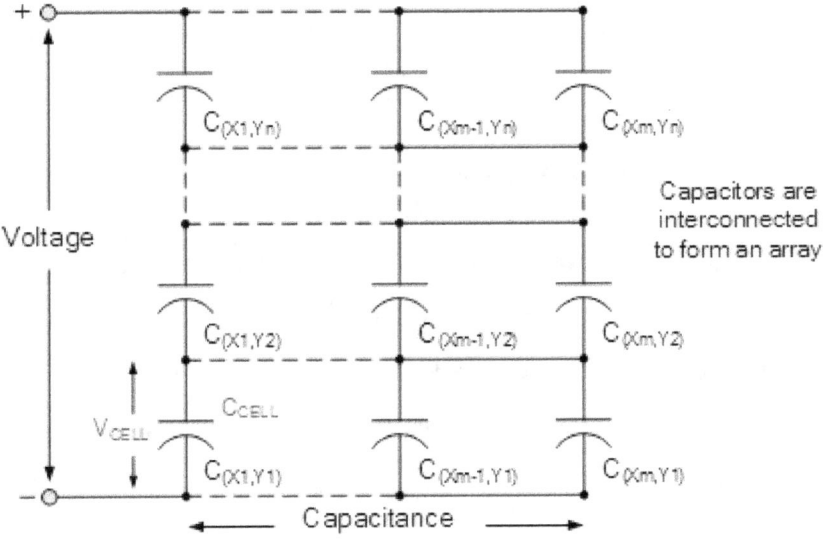

Where: V_{CELL} is the voltage of one cell, and C_{CELL} is the capacitance of one cell.

As the voltage of each capacitor cell is about 3.0 volts, connecting more capacitor cells together in series will increase the voltage. While connecting more capacitor cells in parallel will increase its capacitance. Then we can define the total voltage and total capacitance of a ultracapacitor bank as:

$$\text{Voltage, } V = V_{CELL} \times N$$

$$\text{Capacitance, } C = C_{CELL} \times \left[\frac{M}{N}\right]$$

Where: M is the number of columns and N is the number of rows. Note also that like batteries, ultracapacitor and supercapacitors have a defined polarity with the positive terminal marked on the capacitor body.

Ultracapacitors Example No1

A 5.5 volt, 1.5 farad ultracapacitor is required as an energy storage backup device for an electronic circuit. If the ultracapacitor is to be made from individual 2.75v, 0.5F cells, calculate the number of cells required and the layout of the array.

$$V = V_{CELL} \times N$$

$$\therefore N = \frac{V}{V_{CELL}} = \frac{5.5}{2.75} = 2$$

The array will therefore have two capacitor cells of 2.75v each connected in series to provide the required 5.5v.

$$C = C_{CELL} \times \left[\frac{M}{N}\right]$$

$$\therefore M = \frac{C \times N}{C_{CELL}} = \frac{1.5 \times 2}{0.5} = 6$$

Then the array will have a total of six individual columns, consisting of two rows of six thereby forming an ultracapacitor with a 6 x 2 array as shown.

6×2 Ultracapacitor Array

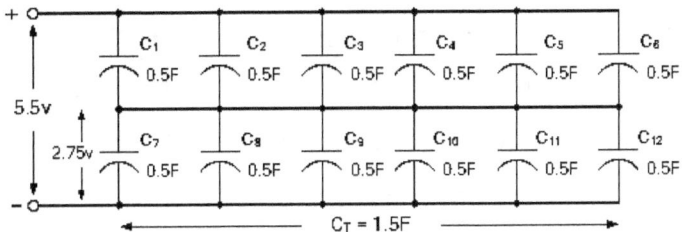

Ultracapacitor Energy

An ultracapacitor is an energy storage device which integrates all capacitors. Electrical energy is stored as charge in the electric field between its plates, resulting in a potential difference, i.e. a voltage, between the two plates.

Electrical energy is stored between the ultracapacitor's plates during charging (current flowing through the ultracapacitor from the connected supply). Once the ultracapacitor is charged, current from the supply ceases and the ultracapacitor's terminal voltage equals the supply voltage.

As a result, a charged ultracapacitor will store this electrical energy even when disconnected from the power supply, acting as an energy storage device. The ultracapacitor converts this stored energy into electrical energy to supply the connected load when discharging (current flowing out).

The amount of energy stored in an ultracapacitor is proportional to the capacitance value of the capacitor, so it doesn't use any energy itself but instead stores and releases electricity as needed. As was mentioned earlier, the relationship between the amount of energy stored and the capacitance C and the square of the voltage V across its terminals gives.

$$E = \frac{1}{2}CV^2 = \frac{CV^2}{2} = \frac{QV}{2} = \frac{Q^2}{2C} \text{ Joules}$$

Where: E is the energy stored in joules. Then for our ultracapacitor example above, the amount of energy stored by the array is given as:

$$E = \frac{1}{2}CV^2 = \frac{1.5 \times 5.5^2}{2} = 22.7 \text{ Joules}$$

Then the maximum amount of energy that can be stored by our ultracapacitor is 22.7 joules, which was originally supplied by the 5.5 volt charging supply. This stored energy remains available as charge in the electrolyte dielectric and when connected to a load, the ultracapacitors entire 22.69 joules of energy is made available as an electric current. Obviously, when the ultracapacitor is fully discharged, the stored energy is zero.

It has been observed that an ideal ultracapacitor would not consume or dissipate energy, but rather would take power from an external charging circuit, store energy in its electrolyte field, and then return this stored energy when delivering power to a load.

In our simple example above, the energy stored by the ultracapacitor was about 23 joules, but with large capacitance values and higher voltage ratings, the energy density of ultracapacitors can be very large making them ideal as energy storage devices.

In fact, ultracapacitors with ratings into the thousands of farads and hundreds of volts are now being used in hybrid electric vehicles (including Formula 1) as solid-state energy storage devices for regenerative braking systems as they can quickly giving out and receiving energy during braking and accelerating afterwards. Ultra and super-capacitors are also used in renewable energy systems to replace lead acid batteries.

An ultracapacitor is a type of electrochemical device that stores charge electrostatically. It consists of two porous electrodes, typically made of activated carbon and submerged in an electrolyte solution. In actuality, this configuration produces two series-connected capacitors, one at each carbon electrode.

Batteries cannot match the power density of ultracapacitors, which are available with capacitances in the hundreds of farads all within a very small physical size.

To provide any useful voltage, however, more than one capacitor must be connected in series and parallel configurations because an ultracapacitor's voltage rating is typically less than 3 volts.

In fact, ultracapacitors are categorized as an ultracapacitor battery because they can be used as energy storage devices that are similar to batteries.

Ultracapacitors, however, can produce much higher power densities in a shorter amount of time than batteries can. Due to their capacity to quickly discharge high voltages and then be recharged once more, ultracapacitors are now widely used in both fuel cell- and gasoline-powered hybrid vehicles.

Peak power demands and transient variations in load conditions can be greatly reduced by combining ultracapacitors with conventional fuel cells and automotive batteries.

www.ingramcontent.com/pod-product-compliance
Lightning Source LLC
Chambersburg PA
CBHW070605220526
45467CB00003B/1304